MCWP 2-6

Counterintelligence

U.S. Marine Corps

5 September 2000

PCN 143 000084 00

To Our Readers

Changes: Readers of this publication are encouraged to submit suggestions and changes that will improve it. Recommendations may be sent directly to Commanding General, Marine Corps Combat Development Command, Doctrine Division (C 42), 3300 Russell Road, Suite 318A, Quantico, VA 22134-5021 or by fax to 703-784-2917 (DSN 278-2917) or by E-mail to **morgannc@mccdc.usmc.mil**. Recommendations should include the following information:

- Location of change
  - Publication number and title
  - Current page number
  - Paragraph number (if applicable)
  - Line number
  - Figure or table number (if applicable)
- Nature of change
  - Add, delete
  - Proposed new text, preferably double-spaced and typewritten
- Justification and/or source of change

Additional copies: A printed copy of this publication may be obtained from Marine Corps Logistics Base, Albany, GA 31704-5001, by following the instructions in MCBul 5600, *Marine Corps Doctrinal Publications Status.* An electronic copy may be obtained from the Doctrine Division, MCCDC, world wide web home page which is found at the following universal reference locator: **http://www.doctrine.usmc.mil**.

Unless otherwise stated, whenever the masculine gender is used, both men and women are included.

MCCDC (C 42)
13 Jul 2004

E R R A T U M

to

MCWP 2-6

COUNTERINTELLIGENCE

1. Change the publication short title to read "MCWP 2-6" (vice MCWP 2-14).

PCN 143 000084 81

DEPARTMENT OF THE NAVY
Headquarters United States Marine Corps
Washington, D.C. 20380-1775

5 September 2000

FOREWORD

Marine Corps Doctrinal Publication (MCDP) 2, *Intelligence*, and Marine Corps Warfighting Publication (MCWP) 2-1, *Intelligence Operations*, provide the doctrine and higher order tactics, techniques, and procedures for intelligence operations. MCWP 2-14, *Counterintelligence*, complements and expands on this information by detailing doctrine, tactics, techniques, and procedures for the conduct of counterintelligence (CI) operations in support of the Marine air-ground task force (MAGTF). The primary target audience of this publication is intelligence personnel responsible for the planning and execution of CI operations. Commanders, planners, and other personnel who use the results from CI operations or provide support to them should also read this publication.

MCWP 2-14 describes aspects of CI operations across the spectrum of MAGTF, naval, joint and multinational operations, including doctrinal fundamentals, equipment, command and control, communications and information systems support, planning, execution, security, and training. MCWP 2-14 provides the information needed by Marines to understand, plan, and execute CI operations in support of the MAGTF across the spectrum of conflict.

MCWP 2-14 supersedes FMFM 3-25, *Counterintelligence*, dated 22 September 1992.

Reviewed and approved this date.

BY DIRECTION OF THE COMMANDANT OF THE MARINE CORPS

B. B. KNUTSON, JR.
Lieutenant General, U.S. Marine Corps
Commanding General
Marine Corps Combat Development Command

DISTRIBUTION: 14300008400

COUNTERINTELLIGENCE

Table of Contents

CHAPTER 1. DOCTRINAL FUNDAMENTALS

Intelligence strives to accomplish two objectives. First, it provides accurate, timely, and relevant knowledge about the enemy (or potential enemy) and the surrounding environment. The primary objective of intelligence is to support decisionmaking by reducing uncertainty about the hostile situation to a reasonable level, recognizing that the fog of war renders anything close to absolute certainty impossible. The second intelligence objective assists in protecting friendly forces through counterintelligence (CI). CI includes active and passive measures intended to deny the enemy valuable information about the friendly situation. CI includes activities related to countering hostile espionage, subversion, and terrorism. CI directly supports force protection operations by helping the commander deny intelligence to the enemy and plan appropriate security measures. The two intelligence objectives demonstrate that intelligence possesses positive—or exploitative—and protective elements. It uncovers conditions that can be exploited and simultaneously provides warning of enemy actions. Thus, intelligence provides the basis for our own actions, both offensive and defensive. Identifying, planning, and implementing MAGTF operations and measures are the main focus of this publication.

1001. OBJECTIVE

CI—Information gathered and activities conducted to protect against espionage, other intelligence activities, sabotage, or assassinations conducted by or on behalf of foreign governments or elements thereof, foreign organizations, or foreign persons, or international terrorist activities. (Joint Publication [JP] 1-02)

The principal objective of CI is to assist with protecting friendly forces. CI is the intelligence function concerned with identifying and counteracting the threat posed by hostile intelligence capabilities and by organizations or individuals engaged in espionage, sabotage, subversion or terrorism. CI enhances command security by denying an adversary information that might be used to conduct effective operations against friendly forces and to protect the command by identifying and neutralizing espionage, sabotage, subversion or terrorism efforts. CI provides critical intelligence support to command force protection efforts by helping identify potential threats, threat capabilities, and planned intentions to friendly operations while helping deceive the adversary as to friendly capabilities, vulnerabilities, and intentions. Physical security reduces vulnerability. Operations security reduces exposure. Combating terrorism makes us a less lucrative target. CI increases uncertainty for the enemy, thereby making a significant contribution to the success of friendly operations. CI also identifies friendly vulnerabilities, evaluates security measures, and assists with implementing appropriate security plans. The integration of intelligence, CI, and operations culminates in a cohesive unit force protection program.

1002. BASIC CONSIDERATIONS FOR CI ACTIVITIES

Hostile Objectives

Adversaries can be expected to use every available means to impede our forces with their efforts directed towards intelligence, espionage, sabotage, subversion, and terrorist operations. Hostile intelligence collection activities are directed toward obtaining detailed knowledge of our forces and their capabilities, limitations, centers of gravity, vulnerabilities, intentions, and probable courses of action. These activities also obtain information concerning the area of operations including weather, terrain, and hydrography.

Adversarial Advantage

Adversary knowledge of friendly operations concentrates efforts on preparing the objective for defense, attacking friendly staging areas, and disrupting the operation through espionage, sabotage, terrorism, and subversive activities. CI is essential to the security of our forces—commencing with routine garrison operations, to the inception of planning, and until the operation is complete—to deny the enemy advantage and manipulate understanding us.

Hostile Espionage Activities

Foreign intelligence services (FIS) capabilities must be accurately assessed. FIS should be assumed to be at least as effective as our own. An adversary should always be given the benefit of the doubt in collecting information and producing intelligence on friendly operations. FIS do not normally develop vital intelligence by obtaining one all-revealing fact. Most worthwhile intelligence is the product of the assembly, comparison, and interpretation of many small and seemingly insignificant items of information.

Enemy Sabotage Activities

Sabotage is an act or acts with intent to injure, interfere with or obstruct the national defense of a country by willfully injuring or destroying, or attempting to injure or destroy, any national defense or war material, premises or utilities including human and natural resources. Immediately prior to the outbreak of hostilities, during combat, and even during military operations other than war (MOOTW), the enemy can be expected to employ sabotage techniques to disrupt friendly operations.

Subversive Activities

Subversive activities are designed and conducted to undermine the authority of friendly forces and/or that of the local government to disrupt friendly activities or to gain aid, comfort, and moral support for the cause of the enemy or hostile force or group. Subversive activity can be directed against individuals, groups, organizations or entire populations. Frequently, subversive activity supports, conceals or provides a favorable environment for espionage, sabotage, and terrorist operations. Subversion is an action designed to undermine the military, economic, psychological, political strength or morale of a regime.

Terrorist Activities

Any personality, organization or installation of political or military significance could be a terrorist target. Terrorists have become adept in the calculated and systematic use or threat of violence in pursuit of their political or ideological goals. Although the tactics and methods of operation of terrorists may vary from group to group, the techniques they employ to dramatize their goals through fear, intimidation, and coercion are similar.

1003. CONCEPTS OF CI AND FORCE PROTECTION

Security—Measures taken by a military unit, an activity or installation to protect itself against all acts designed to, or which may, impair its effectiveness. (JP 1-02)

Historical Services Perspectives

The CI agencies within the four Services have historically demonstrated dramatically different CI areas of emphasis, concepts of operations, and methods of execution.

Naval Criminal Investigative Service and Air Force Office of Special Investigations

The Naval Criminal Investigative Service (NCIS) and the Office of Special Investigations (OSI) have traditionally viewed CI with a strategic focus drawn from their perspectives as, primarily, law enforcement organizations. NCIS is mainly a civilian investigative organization with a chain of command directly from the Secretary of the Navy to the Director, NCIS. The Director, NCIS, has exclusive responsibility for CI policy development and implementation and execution and management of CI programs, with the exception of those combat and combat related CI responsibilities of the Marine Corps. CI activities of NCIS are funded from the Foreign Counterintelligence Program (FCIP). OSI is a field-operating agency of the Air Force. Policy and programmatic oversight rests with the Secretary of the Air Force Office of the Inspector General, not the Director of Intelligence. CI activities of OSI are also funded from the FCIP. Like NCIS, there is no programmatic provision in OSI for tactical intelligence and related activities (TIARA) funding or resources.

Marine Corps and the Army

The Army and Marine Corps maintain CI as a component of their intelligence staffs. The Marine Corps CI orientation is entirely tactical, with funding exclusively within the DON's budget for TIARA. The Army emphasizes both strategic and tactical CI and is supported by a mixture of FCIP and TIARA resources.

Joint

During the development of joint CI doctrine, the issue arose whether CI should fall under the staff cognizance of intelligence or operations because of differences in emphasis and support for CI. Doctrinal evolution has placed the CI of the combatant commands and joint task forces (JTF) command under the joint intelligence staff (J2).

Joint Operations and CI

Exercising command and control of CI assets vary under different circumstances. Military department CI elements are under the command and control of their respective department secretaries to carry out their statutory authorities and responsibilities. However, combatant commanders may choose to exercise staff coordination authority over military department CI elements deployed within their area of responsibility. Staff coordination authority is intended to encompass deconfliction of CI activities and assurance of unity of effort in attaining the military department secretaries' and combatant commanders' CI objectives. If a military operation plan or operation order so specifies, a combatant commander or JTF commander may, on National Command Authority-directed execution, assume operational control of military department CI elements assigned to support the operation for the duration of the operation, including predeployment, deployment, and redeployment phases. Under this authority, Service CI elements are under the combatant commander's authority. MAGTF CI elements, however, are under the operational control of the MAGTF unless otherwise specified. Law enforcement and CI investigations and attendant matters carried out by CI elements remains solely a military department's administrative responsibility.

CI and Intelligence

CI, like intelligence matters, is a command responsibility. In preparing for operations, units must develop a CI plan and implement appropriate CI measures to protect themselves from potential threats. CI is integrated into the overall intelligence effort to identify and counter an adversary's intelligence efforts. Failure to adequately plan for and implement CI operations and measures may result in serious damage to the MAGTF or supported unit. Continuing attention to CI and effective intelligence and operations integration is thus required at all levels of command, from the MAGTF commander to the individual Marine.

CI and Force Protection

Force protection—A security program designed to protect soldiers, civilian employees, family members, facilities, and equipment, in all locations and situations, accomplished through planned and integrated application of combating terrorism, physical security, operations security, personal protective services, and supported by intelligence, CI, and other security programs. (JP 1-02)

Force protection is a responsibility of command. An operations function, force protection is under the staff cognizance of the unit operations officer. CI is a significant contributor to the command's overall force protection effort. Security is a matter of vulnerability and threat assessment with effective risk management. CI helps identify the hostile intelligence threat, assists in determining friendly vulnerabilities to it, and aids with the development of friendly measures that can lessen or negate these. The commander weighs the importance of intelligence and CI to be used as a tool in risk management. Marine Corps CI elements provide unique force protection capabilities through both active and passive CI measures and human resource intelligence (HUMINT) support. Often, CI elements can provide unique intelligence support to the commander's estimate of the situation and situation development (e.g., providing an assessment of the mood of the area of operations, allowing us to feel the pulse of an incident as it develops). CI also provides critical support to the command's overall intelligence efforts by providing indications and warning (I&W) of potential attack and support to targeting and combat assessment efforts.

1004. MAGTF CI OPERATIONS

Responsibilities

The unit intelligence officer plans, implements, and supervises the CI effort for the commander. The G-2/S-2 may have access to or request support from MAGTF CI units and specialists to assist in developing CI estimates and plans. Members of the command are involved in executing the CI plan and implementing appropriate CI measures. Key participants in this process and their specific responsibilities are—

- Unit security manager (generally the chief of staff or executive officer, but often the unit's intelligence officer)—overall integration and effectiveness of unit security practices.
- G-3/S-3—force protection, operations security (OPSEC), counterreconnaissance, and deception.
- G-6/S-6—communications and information systems security.
- G-1/S-1—information and personnel security.
- Headquarters commandant—physical security of unit command post and echelons.

CI Process

The CI process at all levels is conducted by using a standard methodology that consists of four steps: develop a CI estimate, conduct CI survey(s), develop the CI plan, and conduct CI operations and assist with implementation of CI measures. Figure 1-1 summarizes the CI process.

The CI Estimate

Included in CI estimates are known factors on location, disposition, composition, strength, activities, capabilities, weaknesses, and other pertinent information. CI estimates also provide conclusions concerning probable courses of action and future activities of these organizations, effects of those activities on friendly courses of action, and effectiveness of friendly force CI measures. Comprehensive CI estimates are normally prepared by senior echelon commands. Within the MAGTF, intelligence and CI analysts of the MAGTF CE, intelligence battalion (intel bn), and its CI/HUMINT company/detachment will normally prepare a tailored CI estimate that addresses threats to the MAGTF by using an IPB methodology that is focused on CI factors and the CI threat. However, each level of command must conduct its own evaluation to determine which adversary's capabilities identified in the MAGTF CI estimate represent a threat to their particular unit. The CI estimate must be updated on a regular basis, and the revised

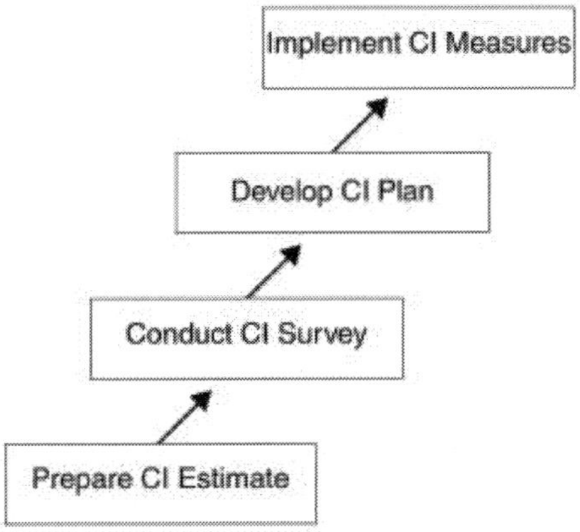

Figure 1-1. The CI Process.

CI estimates provide information on enemy intelligence, sabotage, subversive, and terrorist organizations relevant to the current mission, situation and area of operations.

estimate or appropriate CI warning reports must be disseminated to units involved in the operation.

The CI Survey

The CI survey assesses a unit's security posture against the threats detailed in the CI estimate. The CI survey should identify vulnerabilities to specific hostile intelligence, espionage, sabotage, subversion or terrorist capabilities and provide recommendations on how to eliminate or minimize these vulnerabilities. The survey should be as detailed as possible. During the planning phase of an operation, it may be possible to do a formal, written survey. In a time-compressed situation, the survey will likely result from a brief discussion between the appropriate intelligence, CI, operations, communications, and security personnel. It is critical that the survey look forward in both space and time to support the development of the CI measures necessary to protect the unit as it carries out successive phases of the operation; the survey makes recommendations to improve the CI posture of the command both now and in the future.

The CI Plan

As with all plans, CI plans must be continually updated to ensure they are current and support both ongoing and future operations.

The CI plan details the activities and operations that the command uses to counter hostile intelligence, sabotage, subversion, and terrorist threats. It includes procedures for detecting and monitoring the activities of hostile intelligence and terrorist organizations and directs the implementation of active and passive measures that are intended to protect the force from these activities. The CI plan is based on the threats identified in the CI estimate and the vulnerabilities detected by the CI survey. The intel bn commander, as the intelligence support coordinator (ISC), assisted by CI/HUMINT company, the production and analysis (P&A) cell officer in charge (OIC), and the MEF staff CI officer, will normally prepare a detailed, comprehensive CI plan that addresses the MEF and is integrated with CI plans of the JTF and other pertinent forces. Included in the MAGTF CI plan are details of the employment of dedicated CI capabilities and the conduct of specialized CI operations intended to detect and neutralize or eliminate specific threats. Plans of subordinate MAGTF elements closely follow the MAGTF plan, normally adding only security measures that are applicable to their specific units.

CI Execution

One of the most highly effective tools of the CI collection activity is the counterintelligence force protection source operations (CFSO).

An understanding of the interest and capability of adversarial intelligence organizations to collect information on evolving U.S. technologies is critical to developing appropriate countermeasures. CI personnel can readily obtain information from other national intelligence and security organizations because of its unique liaison arrangements. That capability not only supports the analytical efforts of the national agencies and intelligence centers, but also give an added dimension to I&W. The CFSO provides commanders with a collection and production capability to protect their forces without resorting to complex national coordination procedures. The role of CI is even greater as U.S. military operations increasingly rely upon cooperation and support of our allies. CI personnel can assess the capabilities, effectiveness, organization, and methods of operation of allied intelligence services as well as the effectiveness of their security procedures and ability to support or to detract from the U.S. effort.

1005. CI MEASURES

CI measures—both active and passive—encompass a range of activities designed to protect against hostile intelligence, espionage, sabotage, subversion, and terrorism threats.

CI analysis develops threat assessments that assist decisionmakers in determining the threat posed to their plans, strategies, resources, programs, operations and systems by foreign intelligence activity.

Active Measures

Active CI measures are those designed to neutralize the multi-discipline intelligence effort (all disciplines used to collect intelligence such as HUMINT, signals intelligence [SIGINT], and imagery intelligence [IMINT]) and hostile efforts toward sabotage, subversion, and terrorism. Active CI measures include counterespionage, countersabotage, countersubversion, counterterrorism, counterreconnaissance, concealment, and deception operations and vary with the mission and capabilities of the unit.

Passive Measures

Passive CI measures are designed to conceal and deny information to the enemy, protect personnel from subversion and terrorism, and protect installations and material against sabotage. Measures include security of classified material, personnel security, physical security, security education, communications security, data security, electromagnetic emission security, censorship, camouflage, concealment, light, and security discipline. Passive measures are readily standardized in the unit's standing operating procedures (SOPs) regardless of the unit's mission.

Types of CI Measures

The three general CI measures are denial, detection, and deception. Frequently, the measures applied to accomplish one of these purposes contribute to the others.

Denial Measures

Denial measures are applied to prevent the enemy from gaining access to classified and sensitive information, subverting personnel, and penetrating the physical security barriers established at command posts and echelons, facilities, and installations. Counterreconnaissance is one example of a denial measure that may be used.

Detection Measures

Detection measures are used to expose and to neutralize enemy efforts directed toward intelligence collection, sabotage, subversion, and terrorism. MAGTF units detect or aid in the detection of these enemy efforts by collecting, analyzing, and reporting information on enemy activities that may indicate an intelligence effort by establishing checkpoints to control the movement of personnel within or through their areas of responsibility and by evacuations of possible enemy agents and materials to higher echelons for interrogation and exploitation. Other detection measures, usually accomplished by specialists, include document translation and analysis, screening, interrogation, and offensive and defensive CI activities.

Deception Measures

Control of deception operations should be at the highest level of command likely to be significantly affected by the enemy's reactions.

Deception measures mislead or otherwise confuse the enemy concerning our capabilities, centers of gravity, vulnerabilities, plans, and intentions. Deception measures may include feints, ruses, demonstrations, and the provision of false information to the enemy. Deception measures depend on effective command security for success. Special precautions must be taken ensuring there is no leakage of friendly force information during the planning or execution of an operation. When enemy intelligence activities are identified, consideration must be given to the potential for using that activity in support of deception measures. The potential threat posed by the enemy must be weighed against the potential intelligence benefits of continued exploitation of the enemy's intelligence system versus its destruction or other degradation

1006. CI SUPPORT TO OPERATIONS

MAGTF CI support to operations normally falls within one of the following two categories: support to military security or civil security.

Support to Military Security

Military security encompasses all measures taken by a command to protect itself from sabotage, terrorism, and subversion and to deny information to the enemy. MAGTF units emphasize protection of airfields and other major installations and the defeat of hostile target acquisition efforts. Typical measures include OPSEC, counterreconnaissance, countersigns, passwords, and restrictions on access to selected areas and installations.

Support to Operations Security

OPSEC is the functional responsibility of the operations officer (G-3/S-3).

To be effective, OPSEC vulnerabilities must be determined and countermeasures implemented commencing with operational planning and continuing through completion of any operation. Commanders must determine what essential elements of friendly information (EEFI) and operations must be protected, what OPSEC measures to implement, when to implement them, and what level of risk they are willing to accept. Commanders, staffs, and individuals at all echelons of command are responsible for developing and implementing an effective OPSEC program.

OPSEC denies the enemy prior knowledge of EEFI regarding command activities, plans, operations, strengths, vulnerabilities, and intentions. The enemy collects this information through a variety of means—human, electronic, photographic, etc. To effectively counter this threat, commanders must have access to timely, reliable, and accurate intelligence on enemy intelligence capabilities and operations.

Support to Information Security and C2 Protect

CI supports commanders' OPSEC programs by providing assessments of friendly vulnerabilities; briefings on enemy threats of espionage, sabotage, subversion, and terrorism; and assistance in establishing safeguards and countermeasures against these threats.

The INFOSEC program—a responsibility of the unit's security manager—includes a proper security classification determination being made with applicable security regulations and the proper protection being afforded to the material throughout its life cycle. These measures include proper preparation, reproduction or manufacturing, storage, use, and destruction. Failure to comply with required INFOSEC measures exposes sensitive information to potential compromise. The rapid advancement of the microprocessor and the maturity of computer age technologies have presented a significant new area of exposure that leaves us particularly vulnerable. While these advances provide new capabilities and opportunities, they also create new vulnerabilities to be exploited. Evolving information operations (IO) concepts and doctrine recognizes the potential and the threats created by this trend. IO include actions taken to affect adversary information while defending one's own information and information systems during both routine peacetime, MOOTW, and combat operations. Command and control protect are defensive measures taken to detect and prevent hostile efforts against our C2 and supporting communications and information systems. The ability to directly influence key decisionmakers through the injection, disruption, manipulation or destruction of information and information means is a powerful tool in the advance of military objectives. Information system vulnerabilities include denial of service, information theft, information replacement or introduction of false data. Defensive measures to provide information assurance include use of secure networks, firewalls, encryption, anti-virus scans to detect malicious code, and proper systems administration to include aggressive auditing. Information protection includes the authenticity, confidentiality, availability, integrity, and non-repudiation of information being handled by anyone involved with C2. It requires proper implementation of appropriate security features such as passwords, authentication or other countermeasures. The criticality for CI in this area is the ability to identify the adversary's potential capability to exploit, deny, degrade or destroy friendly C2 before an attack to counter the attempt. Reporting and tracking of attempted and successful attacks will, through trend analysis, assist in the development of countermeasures.

Information security (INFOSEC) is designed to protect sensitive information from potential unauthorized release or compromise.

Counterreconnaissance

Units may be assigned both reconnaissance and counterreconnaissance responsibilities; these two activities complement each other and are inseparable. Good reconnaissance ensures a certain amount of security, and counterreconnaissance provides a certain amount of reconnaissance information. However, a unit tasked with a reconnaissance mission is not ordinarily given a supplementary counterreconnaissance mission as completing the counterreconnaissance mission generally requires neutralizing the hostile reconnaissance elements, while the primary goal of reconnaissance is collection of information without being detected by the enemy. Counterreconnaissance includes setting up a defensive screen to deny enemy reconnaissance or an offensive screen designed to meet and destroy enemy reconnaissance in combat air operations. Counterair operations may be defined as counterreconnaissance when counterair operations deny or reduce an enemy's capability for visual, photographic or electromagnetic reconnaissance.

One of the most effective CI measures taken by a unit is counterreconnaissance.

Principles of Counterreconnaissance. Counterreconnaissance elements focus on friendly forces being screened. Hostile reconnaissance forces are destroyed or neutralized, and friendly screening forces are echeloned in depth.

Forms of Counterreconnaissance. The defensive screen is protective. It is usually established behind natural obstacles. An offensive screen may be moving or stationary depending on the activities of the friendly force being screened. The offensive screen meets the enemy's reconnaissance forces and neutralizes them. The commander's adoption of a form of counterreconnaissance screen depends on the situation, mission, weather, and terrain; thus the form of counterreconnaissance screen adopted, need not reflect solely the tactical mission of the command. Because there are offensive and defensive screens does not imply a requirement for their employment only in support of a like tactical mission. An offensive screen may well be employed to support a tactical mission of defense, while an attack mission may be supported best by a defensive screen.

Support to Embarkation Security

Embarkation security consists of the special application of military and civil security measures to the embarkation phase that include the movement to the point of embarkation and the actual embarkation. Examples include the screening of civilians employed in the port or airfield, control of contact between troops and civilians, covering or removing tactical markings and other unit designations, and moving to the port or airfield under the cover of darkness.

Support to Civil Security

Civil security operations are generally conducted in coordination with law enforcement, civil affairs, and other appropriate agencies.

Civil security operations include CI measures affecting the civilian population of the area. Typical measures include security screening of civilian labor, imposing curfews and other circulation control measures, and the monitoring of suspect political groups.

CHAPTER 2. COUNTERINTELLIGENCE FUNCTIONS AND SERVICES

2001. COUNTERINTELLIGENCE FUNCTIONS

There are four CI functions: operations; investigations; collection and reporting; and analysis, production, and dissemination (see table 2-1).

CI Operations

Offensive CI operations are the employment of specialized CI techniques and procedures. They are directed against the espionage, sabotage, subversive, and terrorism threat. These operations are planned, coordinated, and conducted by MAGTF CI personnel and include the following operations:

Counterespionage Operations

These operations are designed to detect, destroy, neutralize, exploit or prevent espionage activity. This is accomplished through the identification, penetration, manipulation, deception, and repression of individuals, groups or organizations conducting or suspected of conducting espionage activities.

Countersubversion Operations

These operations are designed to detect, prevent or neutralize the activities of subversive groups. Subversive activity is closely related to and frequently supports, conceals or provides a favorable environment for espionage and sabotage operations. Based on this environment, the countersubversive mission may include offensive measures directed toward the origin of hostile subversive plans and policies.

"...Mogadishu has been one tough nut to crack . . . we are making steady and perceptible progress. From my perspective, one of the most encouraging outgrowths of our efforts in this socially, politically and geographically complex urban environment has been the emergence of tactical HUMINT as the driving force behind operations . . . in-by-9 out-by-5 service on priority intelligence requirements."

". . . been directed against clearly defined targets—there have been remarkably few dry holes. Spared the long unproductive walks in the sun sometimes associated with the Vietnam Conflict. The troops have remained alert, tactically disciplined and tightly focused. I believe this accounts, in some measure, for our low casualty rate."

". . . it's refreshing to see things in their proper order—INTELLIGENCE DRIVING OPERATIONS . . . instead of operations driving intelligence."

—MajGen Charles E. Wilhelm
Commander, Marine Corps Forces Somalia

Table 2-1. Objectives of CI Functions.

CI Function	Objectives
CI Operations	Determine foreign intentions. Support tactical and strategic perception management operations. Support all-source intelligence and other CI operations. Support planning and military operations.
CI Investigations	Detect, exploit, prevent, or neutralize espionage activities. Detect and resolve incidents of foreign directed sabotage, subversion, sedition, terrorist activities, and assassinations. Document elements of proof for prosecutions. Provide military commanders and policy makers with intelligence and information to use to eliminate security vulnerabilities and improve security postures.
CI Collections and Reporting	Provide indications and warning of security threats to U.S. forces, facilities, and operations. Provide intelligence on threats to forces to support planning and implementation of defensive or offensive countermeasures. Respond to commander's priority intelligence requirements.
CI Analysis, Production, and Dissemination	Provide analysis and assessments of threats to U.S. forces, facilities, and operations. Provide causal analysis of past events to identify vulnerabilities and risks. Identify adversary organizations, personalities, and capabilities posing threats to forces, facilities, and operations.

Countersabotage Operations

These operations require a comprehensive program to penetrate saboteur, partisan or other dissident groups. The goal of the program is to determine sabotage plans and to identify saboteurs, methods of operation, and specific targets, and thus support MAGTF force protection efforts.

Counterterrorism Operations

These operations are planned, coordinated, and conducted to detect, prevent, or neutralize terrorist groups or organizations by determining terrorist plans or intentions. They are also employed to identify terrorists, methods of operation, and specific targets.

Exploitation and Neutralization Operations

These operations are targeted against personalities, organizations, and installations of intelligence or CI interest, which must be seized, exploited or protected. Screening and interrogations are operations designed to identify and apprehend enemy intelligence agents, subversives, terrorists, and saboteurs who attempt to infiltrate friendly lines and operations or conceal themselves among the civilian population.

CI Investigations

CI investigations are investigations concerning personnel, security matters, espionage, sabotage, terrorism, and subversive activities, including defection. A CI investigation is a duly authorized, systematic, detailed examination/inquiry to uncover and report the facts of a matter. Jurisdiction for CI investigations will vary according to which commander is exercising C2 of the force's CI assets. However, CI investigations and attendant matters carried out by a CI element remain part of the administrative responsibilities of the military department to which the specific CI element is subordinate.

Additionally, MAGTF CI elements may conduct investigations of friendly prisoners of war and persons missing (non-hostile) in action cases. The type of activities required in these investigations include collecting information of potential intelligence value on friendly personnel possibly in enemy hands (including debriefings of returned POW/MIA, with emphasis on identifying, locating, and recovering additional personnel) and the collection of information that aids in identifying, locating, and recovering those friendly personnel known or suspected in enemy hands. Initial damage assessments relating to the possible compromise of operational and sensitive material must be included.

CI Collections and Reporting

CI collections and reporting are a significant force multiplier, which are intended to identify actual and potential threats to the command. Collections include the following.

Liaison

Coordination (within authorized jurisdictional limitations) is conducted by CI elements with local intelligence, CI, security and law enforcement organizations/agencies, and civil affairs and psychological operations units where appropriate. Within the DON, NCIS is the element exclusively assigned to maintain liaison with federal law enforcement, security and intelligence agencies on criminal investigative, CI and security matters. NCIS is the primary agency for liaison in these matters with state local and foreign law enforcement, security and intelligence agencies, including those of foreign and U.S. military departments. Following notification of the local NCIS office, MAGTF CI personnel may conduct liaison necessary to accomplish its mission during combat operations. If prior notification of NCIS is not possible, notification will be made at the earliest opportunity.

CI Force Protection Source Operations

CI force protection source operations (CFSO) are overt source collection activities of an expedient nature intended to identify threats to the command in support of the commander's force protection mission.

CI Analysis and Production

Limited initial analysis is conducted by MAGTF CI elements that originally collect and report the information. Detailed analysis occurs as part of the MAGTF's all-source intelligence effort.

2002. COUNTERINTELLIGENCE SERVICES

They enhance the security of the command against espionage, sabotage, subversion, and terrorism. Technical surveillance countermeasures (TSCM) involve the employment of services and techniques designed to locate, identify, and neutralize the effectiveness of hostile technical surveillance activity (see chapter 10).

CI services include CI surveys and vulnerability assessments, evaluations, inspections, training, and technical services.

2003. CI SUPPORT TO THE STRATEGIC, OPERATIONAL, AND TACTICAL LEVELS OF WAR

The levels of war form a hierarchy. Tactical engagements are components of battle, and battles are elements of a campaign. The campaign is but one phase of a strategic design for gaining the objectives of policy. While a clear hierarchy exists, there are no sharp boundaries between levels; they merge and form a continuum. A particular command echelon is not necessarily concerned with only one level of war. A commander's responsibilities within the hierarchy depend on the scale and nature of the operation and may shift up or down as the operation develops (see MCDP 1-1, *Strategy*, and MCDP 1-2, *Campaigning*, for additional information on the levels of war).

CI provides critical support to all three levels of war (see figure 2-1 on page 2-4). While certain activities may crosslevels based on the environment and the nature of the threat; the levels and type of CI support remain fairly distinct. The distinctions are based on the supported commander's intelligence and CI requirements. If the support satisfies national interests and policy objectives, it is part of the strategic level CI support. When the

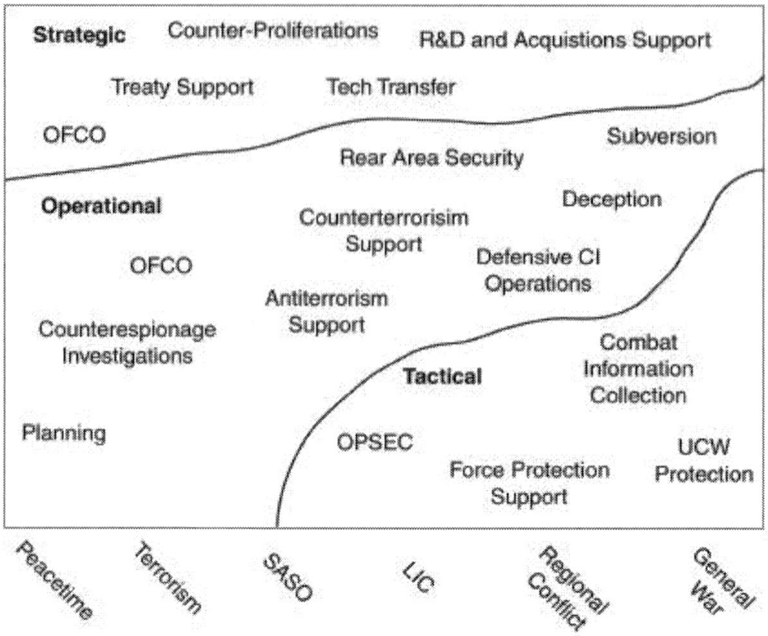

Figure 2-1. Levels of Counterintelligence Support.

objectives and requirements focus on the overall joint force and its operations and sustainment, CI support is at the operational level. Tactical CI support addresses the immediate needs of commanders, particularly maneuver commanders, conducting the battles and engagements, with CI support emphasizing force protection from proximate threats. This publication will focus predominantly on tactical CI support to MAGTF operations.

Strategic CI Support

The strategic level focuses on security objectives, and involves the National Command Authorities, National Security Council, JCS, and Congress. This is a macro perspective in looking at how the instrument of national power is used to satisfy the overarching national objectives, interests, and policies.

Marines support these programs independently and are fully integrated into these programs conducting CI operations, often under the sponsorship of another service or national agencies.

Strategic-level CI helps provide answers to the question, "What threats exist to the national interests and instruments of national power?" Strategic CI primarily supports national-level programs and satisfies requirements across the spectrum of potential threats. Strategic CI emphasizes systems protection, acquisition, proliferation, and strategic level offensive CI operations (OFCO). Although the above programs may also impact on the operational and tactical level, they are generally focused on addressing national-level requirements and support. The impact of these programs should be on national-level decisionmaking with direct linkages supporting combatant commands' strategies and initiatives. National-level agencies centrally manage and control strategic CI support, since the scope of these operations spans across geographic regions and Service or organizational lines.

Operational CI Support

The operational-level commander is the principal supported commander at this level; MAGTF commanders also are key recipients of operational level CI support.

The operational level looks at how to translate the national strategy and objectives into reality through the use of assigned forces. Although strategy—derived from national policies and objectives—defines the nature of the operations, the operational level is responsible for the specific implementation of those strategies. The operational level links strategic security objectives to the tactical decisionmaking and the employment of forces. It is the level that wars are conducted. The operational level looks at the design, organization and integration of strategies, campaigns, and major operations.

Operational CI support, across the joint spectrum of potential employment, is probably the most active area of CI support. It has a major impact on the overall ability of commanders to conduct operations in support of national objectives. Operational CI focuses more on threats to plans and operations, particularly within the context of the wider scope of the campaign, than the more specific scope of the tactical commander. Operational CI focuses on the question of "What are the threats to continuity and the ability to retain the tempo of overall operations?" Operational level CI emphasizes contingency planning, liaison, collections (including CFSO), counterespionage investigations, offensive and defensive CI operations, analysis, production and dissemination of threat related reporting. In contingencies and warfare, many operational CI activities focus on the rear areas, since that is where critical C2, logistics and other sustainment are located.

The combatant command's joint intelligence center and the JTF joint intelligence support element are the principal planners and producers of operational CI support.

Tactical CI Support

Tactical CI is tailored to the needs of the MAGTF and subordinate commanders and addresses their immediate and continuing need for intelligence relating to all manner of threats posed against their forces.

The integration, sustainment, and protection of tactical level forces are of primary concern at this level. The perspective focuses on supporting strategic and operational objectives by implementing and achieving tactical objectives through the use of tactical forces. To a great extent, the tactical level looks at the application of resources (means) applied to achieve national objectives (ends). Responsibility for fighting battles and engagements rests with tactical commanders. Tactical CI support emphasizes direct support to tactical commanders' intelligence and force protection requirements and operations through the identification, neutralization/destruction, and potential exploitation of threats to maneuver forces through the collection of threat related information.

CI personnel conduct vulnerability assessments to look within the command's overall security posture and provide threat assessments to evaluate the infrastructure, capabilities, and intentions of potential threats to the command. They also provide commanders with assessments of the civil population within the AO, and determine their response to the MAGTF presence and actions. The following provides typical examples of CI tactical support.

Non-combatant Evacuation Operations

CI elements deploy to the embassy and other evacuation sites to coordinate and validate the screening of evacuees and assist at the evacuation site to ensure no one attempts to infiltrate with the evacuees.

Peace Operations

CI helps identify and monitor the warring factions and possible third parties to determine potential threats to the peacekeeping/peace enforcement forces.

Humanitarian and Disaster Relief

CI helps identify potential threats to the relief force, providers of aid and assistance, and aid recipients. Often the situation requiring assistance is caused by conflict and can flash with little warning.

Psychological Operations

Psychological operations are normally an activity embedded within other operational activities. CI can assist in gauging the effectiveness of psychological operations and special contingency planning.

Due to the intelligence requirements of commanders in direct contact with hostile forces, the line between CI and HUMINT at the tactical level is blurred almost beyond differentiation. Ground order of battle intelligence is a key area of CI HUMINT collections and production support and seeks to identify enemy forces, dispositions, capabilities, and vulnerabilities. In particular, the identification of threats posed against the MAGTF and the development of countermeasures are key areas CI supports. Intel bn's CI and HUMINT elements are typically task organized into HUMINT support teams (HST). HSTs can satisfy both CI and HUMINT requirements through the collection of threat information from all sources. The teams accomplish this through collection activities including CFSO, overt tactical source HUMINT operations, liaison, interrogation, observation, and debriefings. Threat information in a contingency environment is highly perishable and may have limited utility to anyone other than forces in direct contact.

The standard reporting vehicles are the CI Information Report and the CI SALUTE Report. In addition to this time-sensitive direct support, the intelligence operation center's (IOC) P&A cell is the other key MAGTF producer of tactical CI and HUMINT support.

2004. GARRISON SUPPORT

The primary peacetime/garrison mission of MAGTF CI activities is planning, preparing, and training to accomplish tactical CI functions. A secondary mission is to advise and assist commanders in the planning, coordinating, and implementing of command security and force protection efforts. See chapter 10 for additional information on garrison CI services.

Chapter 3. Organization and Responsibilities

3001. GENERAL

CI, like intelligence, supports all warfighting functions across the spectrum of military operations. Its effective integration within the intelligence effort requires a basic understanding of the national through tactical intelligence organizations. Commanders, through their intelligence officers, depend on coordination and support from many organizations to satisfy their CI and operational requirements.

3002. COMMANDERS AND STAFF PRINCIPALS

Commander

Intelligence and CI are inherent and essential command responsibilities that require the personal involvement of the commander. Commanders at command echelons are responsible for formulating CI plans and implementing CI measures. Commanders must have an understanding of the capabilities and limitations of CI-an understanding of concepts and theory, practical capabilities, limitations and support requirements of their CI personnel, systems, procedures, operations, and products. They must specify CI requirements, focus their efforts and operations, and provide any necessary guidance to ensure timely and useful products and support. CI activities and measures help the commander shape the battlefield for a decisive action. CI measures support effective command security and force protection operations.

While the intelligence officer advises, plans, and implements command CI activities, the commander ultimately determines the effectiveness of the CI effort.

Intelligence Officer

The intelligence officer (G-2/S-2) manages these efforts for the commander, acting as principal advisor on intelligence and CI, and implement activities that carry out the commander's responsibilities. The commander relies on the intelligence officer to provide the necessary information and intelligence on the weather, terrain, and enemy capabilities, status, and intentions. The intelligence officer is a full participant in the commander's decisionmaking process, ensuring that intelligence and CI are effectively used throughout the command during all phases of mission planning and execution.

Through the intelligence operations plan and supporting intelligence, CI and reconnaissance and surveillance plans, the G-2/S-2 validates and plans IRs, coordinates intelligence priorities, integrates collection, production and dissemination activities, allocates resources, assigns specific intelligence and reconnaissance missions to subordinate elements, and supervises the CI and overall intelligence and reconnaissance efforts.

The commander directs the intelligence and CI effort. The intelligence officer has staff responsibility for intelligence and intelligence operations, including CI.

The G-2/S-2's CI responsibilities parallel basic intelligence responsibilities and include—

- Facilitate understanding and use of CI in the planning and execution of operations.
- Use CI to support situation development and the commander's estimate of the situation through the identification of enemy capabilities, strengths, and vulnerabilities as well as opportunities and limitations presented by the environment.
- Provide necessary CI support to command security and force protection operations.
- Assist the commander in developing priority intelligence requirements and supporting CI requirements.
- Develop and answer outstanding MEF and subordinate units' PIRs and IRs by planning, directing, integrating, and supervising organic CI and multi-discipline MEF and supporting intelligence operations.
- Prepare appropriate CI and other intelligence and reconnaissance plans and orders for the MEF and review and coordinate the CI and all-source intelligence plans of JTFs, theaters, and other organizations.
- Supervise the integration of CI in the development and dissemination of all-source intelligence products tailored to the unit's mission and concept of operations.
- Submit and coordinate all-source and CI collection, production, and dissemination requirements beyond the capability of the MEF to satisfy to higher headquarters for JTF, theater, or national CI systems support.
- Evaluate JTF, theater, and national CI and all-source intelligence support and adjusting stated IRs, if necessary.
- Ensure that the command's intelligence and CI requirements are received, understood, and acted on by organic and supporting intelligence assets as part of an integrated, all-source intelligence effort.
- Ensure that CI and other intelligence information is rapidly processed, analyzed, and incorporated where appropriate in all-source intelligence products, and rapidly disseminated to MEF and external units requiring these.
- Monitor the effectiveness of CI activities and the flow of CI products throughout the command and initiate timely corrective action as appropriate.
- Identify and correct deficiencies in CI and other intelligence and reconnaissance personnel and equipment resources.
- Incorporate exercise CI in training exercises to improve MEF individual, collective, and unit readiness.

Operations Officer

The operations officer (G-3/S-3) is the commander's principal staff assistant in matters pertaining to organization, training and tactical operations. In addition to planning, coordinating and supervising the tactical employment of units, the G-3/S-3's principal responsibilities requiring CI support include—

- Planning and coordinating command security (to include operations and signals security).
- Planning and coordinating command force protection operations.

- Recommending missions and, with the intelligence officer, coordinating reconnaissance and counterreconnaissance operations.
- Planning and coordinating electronic warfare and command and control warfare operations and activities (to include electronic protection and C2 protection).

3003. MEG G-2 SECTION AND INTELLIGENCE BATTALION

G-2 Operations Officer

The G-2 operations officer, under the direction of the MEF AC/S G-2, has primary responsibility for intelligence support to the CG and the remainder of the MEF CE in support of current operations and future operations. Specific all-source intelligence and key CI related duties include (see figure 3-1)—

- Coordinating and providing intelligence and CI support to the CG, the G-3 operations section, and the rest of the MEF CE's battlestaff.
- Serving as the G-2 representative to the MEF CE crisis action team.
- Coordinating, providing and supervising intelligence and CI support to the MEF CE current operations center (COC), future operations center (FOC), and force fires.
- Planning, directing, and supervising the Red Cell.

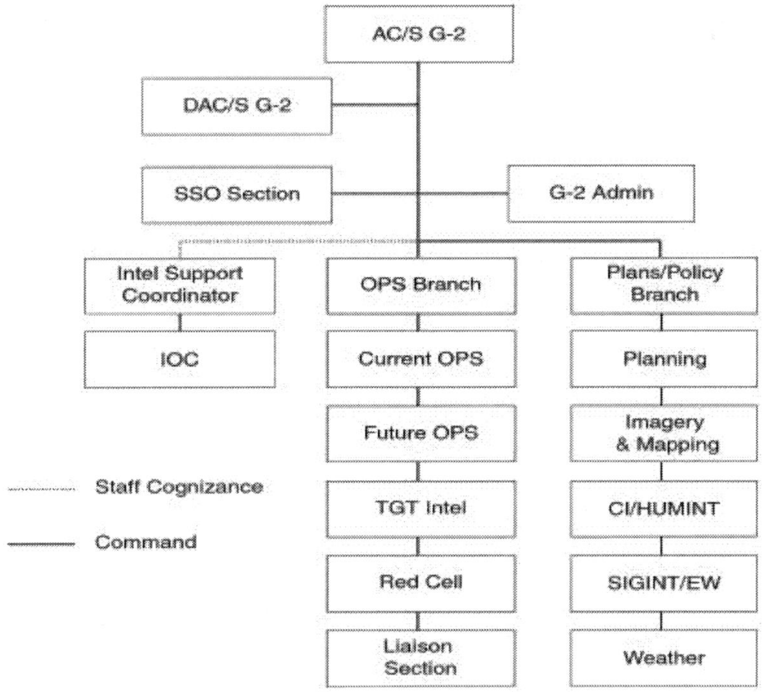

Figure 3-1. MEF G-2 Division Principal Staff Officers and Relationships.

- Providing recommendations on PIR and IR validation, prioritization, and tasking to the AC/S G-2 and the ISC.
- Coordinating and supervising the transition of intelligence and CI planning and operations from G-2 plans to G-2 future operations, and from G-2 future operations to G-2 current operations, to effectively support the MEF's single battle transition process.
- Planning, directing, and supervising MEF liaison teams to external commands (e.g., the JTF and joint functional components headquarters) and intelligence organizations.
- Coordinating with the ISC and MEF MSCs' G-2 operations officers to ensure unity of effort of MEF intelligence and CI operations.
- Providing intelligence and CI input and other support to MEF warning and fragmentary orders and to operations related reporting (e.g., periodic situation reports).
- Coordinating intelligence and CI training for the MEF G-2 section and providing G-2 oversight for and integration of the entire MEF intelligence training program.
- Other intelligence and CI support and tasks as directed by the AC/S G-2.

G-2 Plans Officer

The G-2 plans officer, under the direction of the MEF AC/S G-2, has primary responsibility for intelligence support to the MEF CE's future plans cell. Specific all-source and key CI related duties include (see figure 3-1)—

- Planning the MEF concept of intelligence operations for approval by the AC/S G-2 and subsequent implementation by the ISC based upon the mission, threat, commander's intent, guidance, and concept of operations. This concept of intelligence operations will include a supporting CI concept of operations.
- Leading, coordinating and providing intelligence and CI support to MEF G-5 future plans section.
- Planning and coordinating intelligence and CI support requirements for and the deployment of intelligence elements and resources into the AO.
- Providing recommendations on PIR and IR validation, prioritization, and tasking to the AC/S G-2 and the ISC.
- Coordinating, with the ISC, G-2 development of Annex B (Intelligence) to MEF operations plans (OPLAN), their supporting appendices (such as the initial appendix 3, Counterintelligence, and appendix 5, Human Resources Intelligence), and all intelligence input to other annexes of OPLANs.
- Keeping the G-2 section, other CE staff sections, intelligence liaison personnel, augmentees, and others as appropriate apprised of MEF intelligence and CI planning actions and requirements.
- Identifying requirements and providing recommendations to the G-2 operations officer for MEF intelligence liaison teams to external commands (e.g., the JTF or other components' headquarters) and intelligence agencies.
- Coordinating and developing policies for MEF intelligence, CI and reconnaissance operations.

- Planning, directing, and supervising the MEF G-2's imagery and mapping, CI/HUMINT, SIGINT, and weather sections.
- Accomplishing other intelligence and CI support and tasks as directed by the AC/S G-2.

Intel Bn Commander/Intelligence Support Coordinator

The intel bn commander is responsible for planning and directing, collecting, processing, producing and disseminating intelligence, and providing CI support to the Marine expeditionary force (MEF), MEF MSCs, subordinate MAGTFs, and other commands as directed.

Garrison Operations

In garrison the principal task of the intel bn commander is to organize, train, and equip detachments that support MAGTFs or other designated commands to execute integrated collection, intelligence analysis, production, and dissemination of intelligence products. The composition of intel bn is shown in figure 3-2.

Actual Operations

During operations, the intel bn commander is dual-hatted as the ISC, serving as such under the direct staff cognizance of the MEF AC/S G-2 (see figure 3-1). During garrison operations, many of the tasks listed here are the responsibility of the G-2 operations officer. The intel bn's S-3 section along with the operations center element of the MEF G-2 form the core of the ISC support effort, with planning, direction, and C2 conducted within the IOC's support cell. The ISC is responsible to the MEF AC/S G-2 for the overall planning and execution of MEF all-source intelligence operations. Specific

During garrison operations, many of the tasks listed here are the responsibility of the G-2 operations officer.

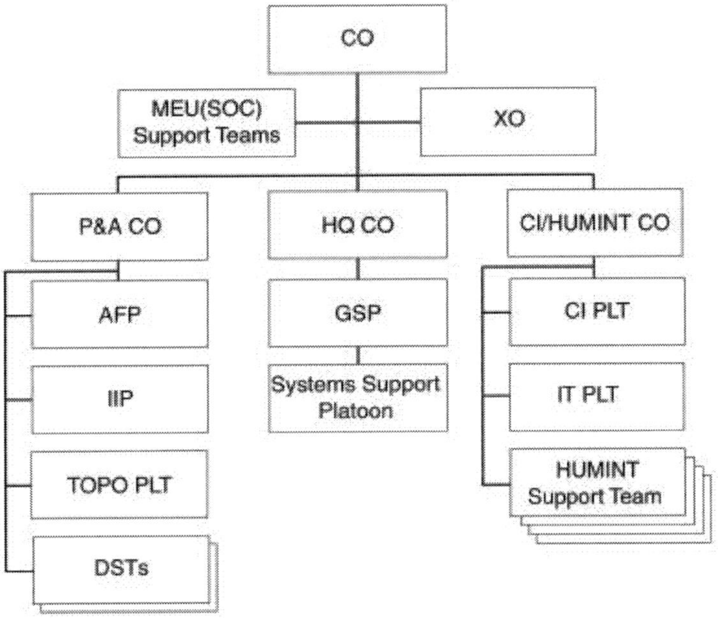

Figure 3-2. Intelligence Battalion.

all-source and key CI responsibilities of the ISC during actual operations include—

- Implementing the concept of intelligence operations and the supporting CI concept of operations developed by the G-2 plans officer and approved by the AC/S G-2.
- Establishing and supervising operation of the MEF intelligence operations center (IOC), including the support cell, the surveillance and reconnaissance cell (SARC), and the P&A cell (see figure 3-3.) Generally the IOC will be co-located with the MEF CE's main command post.
- Establishing and supervising operation of the intel bn's CI/HUMINT company command post.
- Developing, consolidating, validating, and prioritizing recommended PIRs and IRs to support MAGTF planning and operations.
- Planning, developing, integrating, and coordinating MEF intelligence and CI collection, production, and dissemination plans, including the effective organic and external integration and employment of MAGTF CI as well as staff cognizance of MEF SIGINT, IMINT, HUMINT, geographic intelligence (GEOINT), ground remote sensors, ground reconnaissance, and tactical air reconnaissance intelligence collections, production, and dissemination operations.

The ISC is tasked to perform PIR and IR validation and prioritization only during actual operations when the IOC is activated. During routine peacetime operations the PIR/IR validation and prioritization task is the responsibility of the MEF CE's G-2 operations officer.

- Developing, with the G-2 plans officer and G-2 operations officer, and completing Annex B, Intelligence to MEF operations orders (OPORD), supporting appendices (such as appendix 3, CI), and intelligence and CI input to other annexes of OPORDs.
- Planning, developing, integrating, and coordinating intelligence and CI support to the commander's estimate, situation development, indications and warning, force protection, targeting, and combat assessment.
- Managing and fusing the threat (or red) COP/CTP inputs from subordinate units and external commands and intelligence agencies into the MEF CE's threat COP/CTP.
- Providing intelligence and CI support to the MEF CE G-2 section and the MSCs.
- Preparing the intelligence and CI estimates to support G-2 plans.

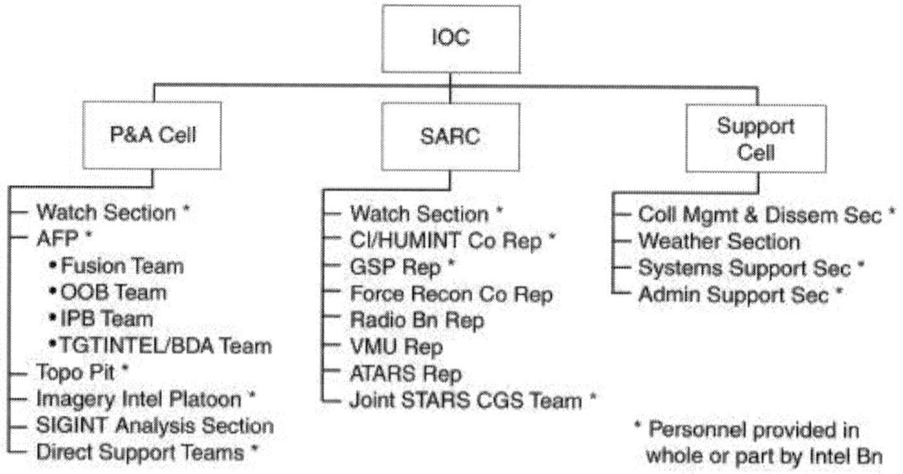

Figure 3-3. Intelligence Operations Center.

- Planning, developing, and coordinating intelligence communications and information systems architecture, including its integration and support of MEF CI and other intelligence and reconnaissance requirements.
- Coordinating and integrating MEF CI and all-source intelligence operations with other service components, JTF joint intelligence support element (JISE) and the joint force J-2 CI/HUMINT staff element (J-2X), theater joint intelligence center (JIC) or joint analysis center (JAC), and national intelligence agencies and operations (e.g., NIMA), including all aspects of intelligence and CI reach-back support.
- Assisting with the evaluation and improvement of MEF CI and all-source intelligence operations.
- Accomplishing other intelligence and CI support and tasks as directed by the AC/S G-2.

(See table 3-1 for a summary of the principal responsibilities of the AC/S G-2's three principal staff subordinate officers.)

CI/HUMINT Officer

During garrison operations the CI/HUMINT Officer (CIHO) is responsible to the G-2 plans officer, in coordination with the intel bn commander, for the planning, direction, and execution of MEF CI/HUMINT operations. In intelligence sections without a CIHO, a designated officer will generally be assigned to perform these tasks. If a HST or other CI/HUMINT Co element is attached, its senior CI/HUMINT officer will serve in this capacity. During action operations, the CIHO's specific duties include—

- Preparing MAGTF CI/HUMINT concept of operations, plans and orders; and directing, coordinating and managing organic and supporting CI/HUMINT operations with the intel bn commander/ISC, G-2 plans officer, IOC's support cell OIC, and the CI/HUMINT company commander.
- Coordinating, planning, supervising, and assisting CI collection requirements and taskings for MAGTF operations with the intel bn commander/ISC, collection management and dissemination officer (CMDO) and CI/HUMINT Co planners.
- Maintaining liaison with other CI and HUMINT agencies.

Table 3-1. AC/S G-2's Principal Subordinate Staff Officers and Their Responsibilities.

ISC	G-2 Ops O	G-2 Plans O
Planning and execution of intel ops to support all MEF IRs	Intelligence support to MEF CE battle-staff and current ops center agencies	Intelligence support to the G-5 future planning team for future planning IRs.
Establish and direct the IOC (P&A Cell, SARC, and Support Cell)	Coordinate and support to higher and adjacent HQs and agencies	Recommends IR validation, prioritization and tasking to AC/S G-2
IR management (collection, production, and dissemination) validation, prioritization, and tasking per AC/S G-2 direction	Recommends IR validation, prioritization, and tasking to AC/S G-2	Establish and direct the G-2 future planning intelligence element
Intel ops command of Intel bn and staff cognizance over SIGINT, CI, HUMINT, MASINT, IMINT, and air-ground recon (includes staff cognizance of designated G-2 elements)	Establish and direct intelligence elements and support to the COC, FOC, Tgt Intel Sec, force fires, Red Cell, and MEF intelligence liaison teams	G-2 section's imagery and mapping, CI/HUMINT, SIGINT, and weather sections (less that under staff cognizance of the ISC)

In intelligence sections without a CIHO, a designated officer will generally be assigned to perform these tasks. If an HST or other CI/HUMINT Co element is attached, its senior CI/HUMINT officer will serve in this capacity.

- Planning for the timely reporting of CI/HUMINT-derived intelligence to MAGTF and external elements and the rapid handling of perishable CI/HUMINT information with the intel bn commander/ISC and CMDO.
- Assisting the intel bn commander/ISC, G-2 plans officer, and G-2 operations officer with preparing and presenting intelligence briefings and reports as required.
- Serving as the principal point of contact between the command and NCIS in matters involving the investigation of actual, potential, or suspected espionage, sabotage, terrorism intelligence, and subversive activities, including defection, ensuring that information about these activities is reported promptly to the nearest NCIS representative. Informing the Criminal Investigation Division (CID) of the Provost Marshal Office (PMO) on criminal matters. These include those of a terrorist nature uncovered in the course of a CI investigation.
- Monitoring command CI/HUMINT MOS training and providing advice and assistance for the maintenance of an effective program.
- Coordinating with the G-2 operations officer, ISC and G-3 to provide CI support to the command's force protection mission, including OPSEC, and deception.
- Providing personnel to jointly man the CAT with CID, NCIS, and if required, civilian law enforcement agents when a crisis action team is established in response to a terrorist or criminal situation,.
- Assuming the role as the Task Force CI Coordinating Authority (TFCICA) when the MAGTF CE is designated as a JTF headquarters.

Collection Management/Dissemination Officer

The CMDO is sourced from the intel bn's S-3 section and is a key subordinate to the intel bn commander/ISC during operations. The CMDO is responsible for formulating detailed intelligence and CI collection requirements (ICRs) and intelligence dissemination requirements (IDR) and tasking and coordinating internal and external operations to satisfy these. The CMDO receives validated PIRs and IRs and direction from the ISC, and then plans and manages the best methods to employ organic and supporting collection and dissemination resources through the intelligence collection and dissemination plans (tabs to Appendix 16, Intelligence Operations Plan, to Annex B), which includes CI collection and dissemination activities. The CMDO is responsible for validating and forwarding national and theater CI and other collection requests from the MEF and MSCs typically using appropriate intelligence and CI tools and TTP. The CMDO is also responsible for coordinating intelligence and CI CIS requirements and maintaining awareness of available CIS connectivity throughout the MAGTF and with key external organizations. During operations the CMDO works within the support cell (see figure 3-3). With the P&A cell OIC, the SARC OIC, G-2 operations officer, CI/HUMINT Co CO, and the MEF G-6, the CMDO is responsible to the ISC for the following CI-related tasks:

- Determining and coordinating the collection effort of PIRs/IRs via organic and supporting CI resources.
- Evaluating the effectiveness of MEF and supporting CI collection and dissemination operations.

- Determining PIRs/IRs and preparing requests for intelligence (RFI) that are beyond organic capabilities and preparing submissions to higher headquarters and external agencies for support.
- Recommending dissemination priorities, development of intelligence and CI reporting criteria, and advising and selecting dissemination means.
- Developing and coordinating CI and all-source intelligence collection plans, coordinating and integrating these with MEF, other components, JTF, theater, and national intelligence collection and dissemination operations.
- Developing and coordinating CI and all-source intelligence dissemination plans and supporting architectures for both voice and data networked communications, and coordinating and integrating these with MEF, other components, JTF, theater, and national intelligence CIS and dissemination operations.
- Monitoring the flow of CI throughout the MAGTF and ensuring that support is delivered to intended recipients in a timely fashion and satisfactorily meets their needs.

Surveillance and Reconnaissance Cell OIC

The SARC OIC is also an immediate subordinate of the ISC and is responsible for supervising the execution of the integrated organic, attached, and direct support intelligence collection and reconnaissance operations (see figure 3-3). The SARC OIC is responsible to the ISC for accomplishing the following specific CI-related responsibilities:

- Coordinating, monitoring, and maintaining the status of all ongoing CI collection operations. This includes—
  - Missions, tasked ICRs, and reporting criteria for collection missions.
  - Locations and times for pertinent fire support control measures.
  - Primary and alternate CIS plans for both routine and time-sensitive requirements, for CI collectors as well as collectors or the SARC and key MEF CE and MSC C2 nodes, to support ongoing C2 of CI collection operations and dissemination of acquired data and intelligence to those needing it via the most expeditious means.
- Conducting detailed CI collection planning and coordination with the MSCs and CI/HUMINT Co planners, with emphasis on ensuring understanding of the collection plan and specified intelligence reporting criteria.
- Ensuring other MAGTF C2 nodes (e.g., the current operations center, force fires, etc.) are apprised of ongoing CI and other intelligence and reconnaissance operations.
- Receiving routine and time-sensitive CI-related reports from deployed collection elements; cross-cueing among intelligence collectors, as appropriate; and the rapid dissemination of CI reports to MAGTF C2 nodes and others per standing PIRs/IRs, intelligence reporting criteria, the dissemination plan, and the current tactical situation.

Production and Analysis Cell OIC

The P&A cell OIC is the third principal subordinate to the ISC, with primary responsibility for managing and supervising the MEF's all-source intelligence and CI processing and production efforts (see figure 3-3 on page 3-6), including aspects of CI production. Key all-source and CI-related responsibilities include—

- Planning, directing, and managing operations of the all-source fusion platoon (to include the fusion, order of battle, IPB, and target intelligence/ battle damage assessment teams), the topographic platoon, the imagery intelligence platoon (IIP), the direct support teams (DST), and other analysis and production elements as directed.
- Coordinating and integrating P&A cell operations, estimates and products with the MEF G-2 section's operations branch and its Red Cell operations and estimates.
- Maintaining all-source-automated intelligence and CI data bases, files, workbooks, country studies and other intelligence studies.
- Planning and maintaining CI, imagery, mapping and topographic resources and other intelligence references.
- Administering, integrating, operating, and maintaining intelligence processing and production systems, both unclassified general service and SCI information systems (e.g., CHATS).
- Analyzing and fusing CI with other intelligence and CI into tailored all-source intelligence products to satisfy all supported commanders' stated or anticipated PIRs and IRs.
- Developing and maintaining current and future intelligence situational, threat, and environmental assessments and target intelligence based upon all-source analysis, interpretation, and integration.
- Managing and fusing the threat (or red) COP/CTP inputs from subordinate units and external commands and intelligence agencies into the MEF CE's threat COP/CTP.

CI/HUMINT Companies

The counterintelligence/HUMINT company (CI/HUMINT Co) is organic to the intel bn within each MEF. CI/HUMINT Co is organized and equipped under tables of organization number 4713 and 4714. It is under the command of the intel bn commander, with any detachments from it under the command of its OIC. The MEF commander exercises command and control of MEF CI/HUMINT operations through the MEF AC/S, G-2, to accomplish the CI/HUMINT mission. The AC/S, G-2 exercises direction of intelligence battalion and the CI/HUMINT company through the intelligence operations officer and the CIHO. The intel bn commander, MEF headquarters group, exercises command and control of the CI/HUMINT Co.

Mission

The mission of the CI/HUMINT Co is to provide CI and HUMINT support to the MEF, other MAGTFs, and other units as directed.

Tasks

- Conduct tactical CI activities and operations, to include CFSO.
- Conduct screening, debriefing, and interrogation of personnel of intelligence/CI interest.
- Direct and supervise intelligence activities conducted within the interrogation facility and the document and material exploitation facility.

- Perform CI and terrorism threat analysis and assist in the preparation of CI and intelligence studies, orders, estimates, and plans.
- Conduct overt HUMINT operations.
- Collect and maintain information designed to identify, locate, and recover captured or missing personnel.
- Debrief friendly personnel recovered from enemy prisoner of war (EPW), hostage or detainee status.
- Translate and exploit captured documents.
- Assist in the conduct of tactical exploitation of captured material and equipment.
- Conduct limited CI investigations during combat or operations other than war.
- Conduct CI surveys and evaluations.
- Conduct TSCM operations.
- Maintain foreign area specialists who provide sociological, economic, cultural and geo-political information about designated countries.

Organization

A CI/HUMINT Co consists of a headquarters section, a CI platoon, an interrogator-translator (IT) platoon and two to five HUMINT Support Teams (HSTs). The CI platoon is organized into a platoon headquarters, four CI teams (CIT), and a TSCM team. The IT platoon is organized into a platoon headquarters and six IT teams (ITT). Figure 3-4 shows the organization of CI/HUMINT Co within I and II MEF (T/O # 4714); figure 3-5 on page 3-12 shows the organization of III MEF's CI/HUMINT Co (T/O # 4713).

Command and Control and Concept of Employment

Command and Control. The CI/HUMINT Co is a subordinate unit of the intel bn, with the intel bn commander maintaining full command of its

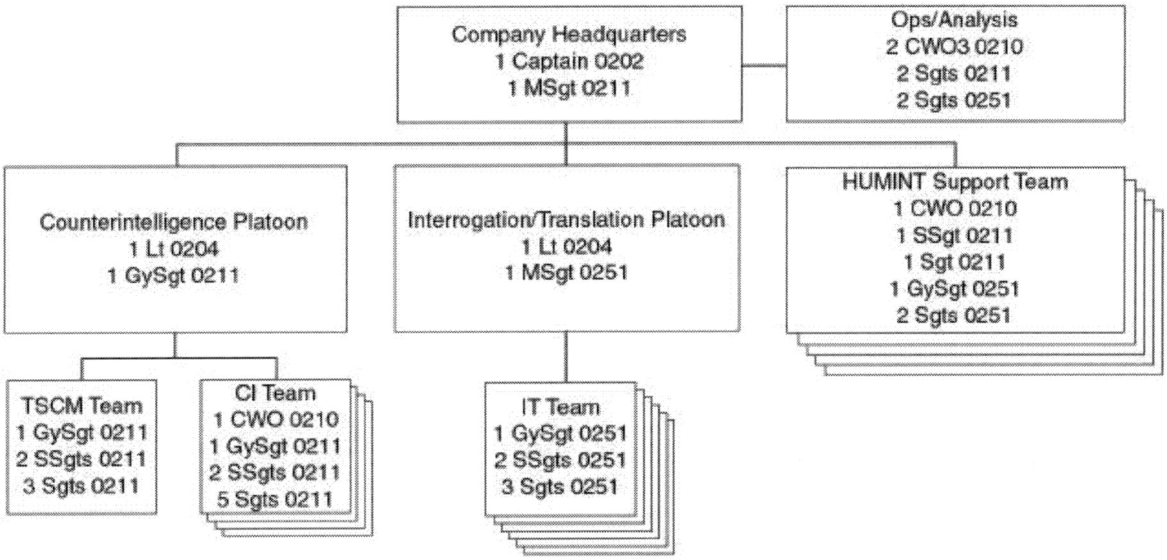

Figure 3-4. CI/HUMINT Company, 1st and 2d Intel Bns, I and II MEF.

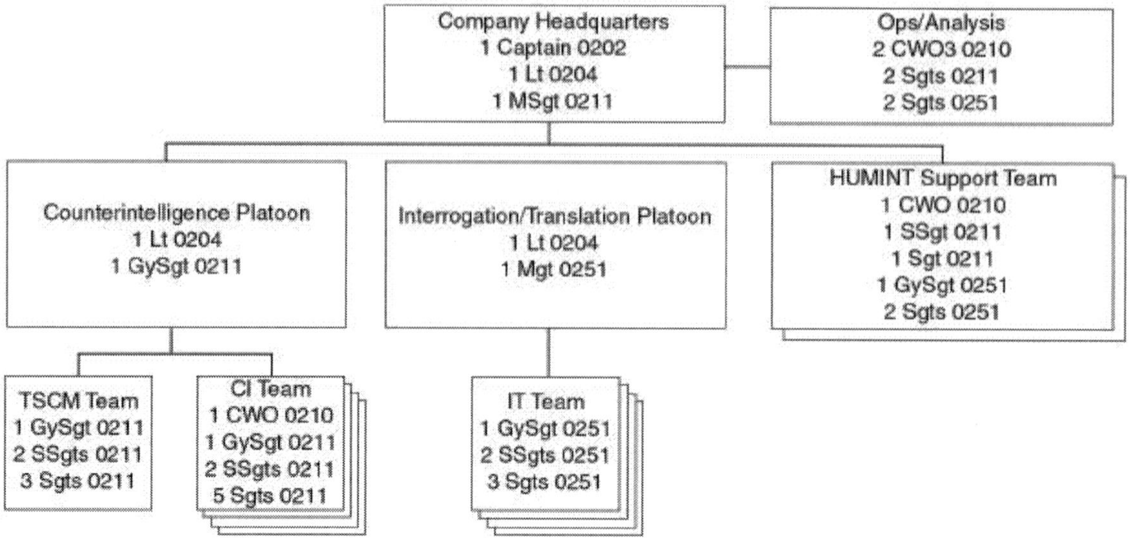

Figure 3-5. CI/HUMINT Company, 3d Intel Bn, III MEF.

operations. The CI/HUMINT Co commander exercises full command authority over subordinate elements, fewer elements that have been detached under a particular task organization. When supporting smaller MAGTFs, CI/HUMINT Co or its detachments will operate under the C2 of either the intel detachment OIC or the supported unit's G-2/S-2. The CI/HUMINT Co is under the command of the intelligence battalion commander. Operational control (OPCON) of intel bn rests with the MEF commander. The MEF commander exercises OPCON through the G-2. Tactical control (TACON) or specified C2 support relationships of CI/HUMINT elements may be provided to MEF subordinate commanders depending upon the situation. However, regardless of the C2 and support relationships, the MEF commander will generally retain technical control of all MEF CI/HUMINT CO elements.

MEF CE Staff Cognizance. The MEF commander will usually exercise C2 over the intel bn elements, to include the CI/HUMINT Co, via the MEF AC/S G-2. The ISC performs this function under the staff cognizance of the AC/S G-2. The intel bn commander/ISC exercises C2 of CI/HUMINT Co via its commanding officer. This allows for the centralized direction and effective integration of CI/HUMINT Co operations within the MEF's broader all-source intelligence concept of operations.

Concept of Employment and C2 Support Relationships. The CI/HUMINT Co combines the MEF's CI and IT capabilities into one organization to provide unity of effort of CI and HUMINT operations in support of MAGTF intelligence and force protection needs. The company is employed per the concept of intelligence support, the CI and HUMINT plans, and the intelligence operations plan developed by the MAGTF G-2/S-2. Subordinate elements of the company may be placed in GS of the MEF, placed in direct support of subordinate commands, or attached to subordinate elements. Additionally, a task-organized detachment will usually be provided to most subordinate MAGTFs and may be used to support joint operations. Figure 3-6 portrays typical notional concept of employment and task organization of CI/HUMINT Co and its elements.

- **General Support.** CI/HUMINT Co will typically operate in GS of the MEF. Under GS, the MEF commander, through the G-2 and the intel bn commander/ISC, determines priorities of intelligence collections and production support, locations of CI support nodes, and CI and all-source intelligence dissemination.
- **Direct Support and Attached.** Depending upon mission, enemy, terrain and weather, troops and support available, time available (METT-T) factors and considerations, the CI/HUMINT Co or task-organized HSTs or other detachments from it may be employed in direct support of or attached to a particular unit or MSC of the MEF. In such cases the scope of the supported commander's C2 authority over assigned CI/HUMINT Co elements will usually be specified to ensure effective support to his operations while allowing the MEF commander to maintain effective C2 of broader intelligence and CI operations.
- **Technical Control.** TECHCON is the performance of specialized or professional service, or the exercise of professional guidance or direction through the establishment of policies and procedures. The nature of the threat CI targets is such that within the MEF, CI TECHCON generally will be retained, coordinated and exercised by the MEF commander via the AC/S G-2 and exercised via the ISC regardless of any other C2 relationships established for the operation. For example, the MEF commander will generally retain CI TECHCON over CI/HUMINT Co elements attached to or placed in direct support of MAGTF subordinate elements.
- **Annex B and the Intelligence Operations Plan.** Specific details regarding C2 relationships over MAGTF CI/HUMINT company

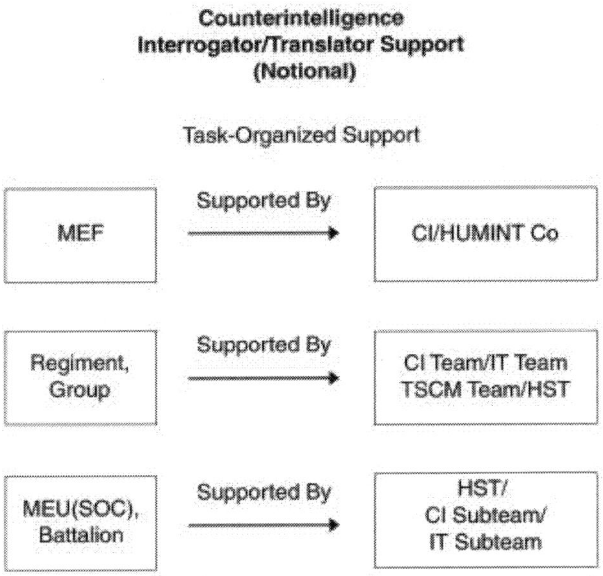

Figure 3-6. CI/HUMINT Company Notional Concept of Employment and Task Organization.

Staff cognizance—The broad responsibility and authority over designated staff functions assigned to a general or executive staff officer (or their subordinate staff officers) in his area of primary interest. These responsibility and authorities can range from coordination within the staff to the assignment or delegation to the staff officer by the commander to exercise his authority for a specified warfighting function or sub-function. Staff cognizance includes the responsibility for effective use of available resources and may include the authority for planning the employment of, organizing, assigning tasks, coordinating, and controlling forces for the accomplishment of assigned missions. Marine Corps orders and doctrine provide the notional staff cognizance for general or executive staff officers, which may be modified by the commander to meet his requirements. (MCWP 6-2, *MAGTF C2*) operations and resources will generally be detailed within the operations plan (OPLAN) or OPORD. This usually will be in one of the following documents: Paragraph 5 to the basic annex B; appendix 3 (CI) to annex B; or appendix 16 (Intelligence Operations Plan) to annex B.

Administrative, Logistics, and Other Support

Administrative. CI/HUMINT Co and its subordinate units are not capable of self-administration. Administrative support is provided by the MEF Headquarters Group (MHG). Administrative support for HSTs and CI/HUMINT Co detachments is provided by the supported unit.

Maintenance and Supply. CI/HUMINT Co and its subordinate units are capable of first echelon maintenance support of organic equipment. Higher maintenance is provided by the MHG or other designated external or supported units.

Transportation. CI/HUMINT has limited organic vehicular transportation support to support Co and subordinate units operations. External transportation support from the MHG, other designated unit or the supported unit is necessary to displace all company elements.

Selected Items of Equipment

TAMCN	Description	Nomenclature	Qty
A03809	Counterintelligence Equipment, Tech Surveillance		1
A0890	Facsimile, Digital, Lightweight	AN/UXC-7	9
A1260	Navigation Set Satellite (PLGR)	AN/PSN-11	25
A2030	Radio Set	AN/PRC-68A	2
A2065	Radio Set	AN/PRC-104B(V)	8
A2070	Radio Set	AN/PRC-119A	23
A2145	Radio Set	AN/VRC-46	5
A2167	Radio Set	AN/VRC-88A	20
D0850	Trailer, Cargo, 3/4-ton, two-wheel	M-101A3	7
D1158	Truck, Utility, Cargo/Troop Carrier 1/4-ton, HMMWV	M-998	35

HUMINT Support Team

The HST is the smallest element to deploy in support of a MAGTF and often serves as the basic building block for CI/HUMINT Co support to subordinate elements of the MAGTF. Specific elements and capabilities provided in the detachment will be based upon the mission of the supported unit, commander's intent, results of the intelligence preparation of the battlespace, the supported unit's concepts of operations and intelligence, and other METT-T factors.

Mission

HSTs support the MAGTF's focus of effort or other designated units, exploit significant HUMINT or CI collection opportunities, or provide tailored support to individual subordinate elements of the MAGTF, in particular elements operating independently from the rest of the MAGTF. For example, a HST is usually attached to the Marine expeditionary unit (special operations capable) MEU (SOC) command element.

Tasks

- Conduct offensive and defensive CI activities, including counterespionage, countersabotage, countersubversion, and counterterrorism in support of tactical units.
- Conduct intelligence collection operations using CFSOs and overt tactical source HUMINT operations.
- Advise commanders concerning support to their force protection, OPSEC, deception, and security programs.
- Assist in the preparation of CI estimates and plans for AO reflected by concept/operation plans.
- Maintain information and CI data bases concerning personalities, organizations, installations, and incidents of CI interest in support of concept/operation plans.
- Collect and maintain information designed to identify, locate, and recover friendly personnel captured, mission (non-hostile), and missing in action.
- Conduct CI debriefings of friendly EPWs who are returned to U.S. control.
- Conduct liaison with unit, JTF, other services, allied, and host nation intelligence and local intelligence, CI, and law enforcement agencies as appropriate.
- Conduct CI investigations about espionage, sabotage, terrorism, subversion, and defection; and other special CI investigations, during combat operations per theater directives.
- Conduct debriefings/interrogation of known or suspected foreign intelligence personnel and agents taken prisoner.
- Maintain foreign language proficiency to support operations.
- Assist in Cl surveys/vulnerability assessments of commands and installations to determine the security measures necessary to protect against espionage, sabotage, subversion, terrorism, and unauthorized disclosure or access to classified material.
- Debrief Marine Corps personnel detained/held hostage by foreign governments or terrorist organizations.

T/Es for CI/HUMINT company had not been finalized by the time of publication of this manual. Refer to the current tables of equipment for accurate current information.

Organization

HSTs combine CI and IT personnel in a single unit to integrate CI and HUMINT collection capabilities. The HST normally consists of one CI officer, two CI enlisted specialists, and three enlisted ITs. Its specific composition, however, will be based upon the mission. HSTs are capable of planning and executing CI/HUMINT operations supporting the designated unit's intelligence and force protection requirements.

Command and Control and Concept of Employment

Command and Control. HSTs are normally attached to the units they support, but they may be placed in direct support. When attached they are OPCON to the supported commander, who exercises C2 via the unit's intelligence officer. Finally, TECHCON of HST operations generally will be retained by the MAGTF commander and exercised via the MAGTF G-2/S-2 and the ISC.

Concept of Employment. HSTs will be employed per the supported commander's concepts of intelligence and operations and the specified C2 relationships.

3004. INDIVIDUAL MARINES

Marines—regardless of rank or military occupational speciality (MOS)—will ensure that their unit's security is not compromised through comprehensive understanding of the unique security vulnerabilities of their operations and functions and the enforcement of necessary personnel, information, operations, and electronic security measures.

3005. MARINE CORPS CI ORGANIZATIONS WITHIN THE SUPPORTING ESTABLISHMENT

The Director of Intelligence is responsible to the CMC with primary responsibility for developing and monitoring Marine Corps CI policy implementation throughout the Marine Corps. The Head, CI/HUMINT Branch of the Intelligence Department, Headquarters U.S. Marine Corps is the principal advisor to the Director of Intelligence for CI/HUMINT matters. The CI/HUMINT Branch performs the following functions:

- Prepares plans, policies, and directives, and formulates controlled CI/HUMINT missions.
- Coordinates with national-level Department of Defense (DOD) and non-DOD agencies on matters dealing with Cl and HUMINT.
- Acts as the CI military occupational specialty (MOS) sponsor (MOS 0204/0210/0211) and IT MOS sponsor (MOS 0251).
- Maintains staff cognizance over CI field units and staff management of training.
- Exercises staff responsibility for HUMINT resources and certain classified/special access programs.
- Coordinates with the NCIS in special investigations and operations.
- Conducts security reviews.
- Reviews reports from the field commands concerning security violations, loss of classified material, and compromises.
- Coordinates the release of information for foreign disclosure.
- Represents the Marine Corps on national level interagency CI committees and subcommittees.

3006. NAVAL COMPONENT ORGANIZATION

N-2 Intelligence Officer

While afloat, the N-2 coordinates activities of the attached NCIS agent to identify threats to the amphibious ready group (ARG). The N-2 coordinates vulnerability/threat assessments and other intelligence and CI in support of ARG intelligence and force protection requirements for the ships. Close coordination between the MAGTF G-2/S-2 and the N-2 ensures consolidation and dissemination of applicable threat related data for the formulation of the commander's estimate. Monitoring and rapid reporting of time sensitive information and intelligence is critical for these deployed commanders as the situation develops. The critical role is the monitoring for indications and warning of impending attack during movement or the deployment of forces ashore. The ARG/MAGTF team continues this interactive relationship through the rapid collection, processing, production, and dissemination of intelligence and CI support of intelligence and force protection requirements.

Attached NCIS Agent

With MAGTF CIHO and other CI elements, the NCIS agent afloat assists the command with the identification of threat and vulnerabilities of the ARG. As a special staff officer under the staff cognizance of N-2, the NCIS agent also has the responsibility for criminal investigation with the ship's master-at-arms and the MAGTF CID officer. During contingency operations, the NCIS agent serves as the principal planner and host for NCIS's surge capability via its Special Contingency Group. The Special Contingency Group is a task-organized group of specially trained NCIS agents prepared and equipped for deployment into tactical situations.

In accordance with DODINST 5240.10, *DOD Counterintelligence Support to Unified and Specified Commands*, each combatant command has a special staff officer within the J-2 with staff cognizance for CI activities within the theater.

3007. JOINT CI ORGANIZATION

CI Staff Officer

The CI Staff Officer is the J-2's primary staff officer responsible for advising, planning, and overseeing theater CI activities. Responsibilities include—

- Advise commander and J-2 on CI investigations, operations, collections, and production activities affecting the command.
- Advise commander on counterdrug, OPSEC, counterterrorism, and antiterrorism activities in the command area of responsibility (AOR).
- Coordinate CI support activities within combatant command's headquarters staff and with component organizations.
- Coordinate the combatant commander's CI requirements with pertinent U.S. CI organizations and U.S. country teams as required.
- Coordinate tasking of CI within AOR and area of interest on implementation of NCA approved/directed action.

- Coordinate with the military services for integrated CI support to research and development and acquisition programs to protect sensitive or critical technologies.
- Ensure significant CI threat information developed within commander's AOR is forwarded to the J-2, other staff officers, and subordinate component commanders.
- Ensure CI support requirements are identified and satisfied during development of command intelligence architecture plans.
- Ensure CI staffing and analytic and production support is integrated into the combatant command's JICs.
- Ensure CI collection, production, and dissemination priorities are integrated into command's intelligence operations plans.

Task Force CI Coordinating Authority

The task force CI coordinating authority (TFCICA) is a JTF headquarters staff officer designated by the combatant commander as the executive agent for CI activities within a JTF's AOR. The TFCICA coordinates and deconflicts with the Defense HUMINT Service's representative to the JTF and the HUMINT Operations Cell (HOC). Together these two staff responsibilities combine to create the JTF headquarters J-2X, which has the task of coordinating and deconflicting CI and HUMINT activity within the JTF AOR. This includes coordination with the U.S. country team and any external attachments and agencies conducting CI and HUMINT activity within the AOR.

3008. NATIONAL LEVEL COUNTERINTELLIGENCE SUPPORT

The Defense Intelligence Agency (DIA) is critical in the planning and establishment of military CI activities.

DIA is the principal DOD organization for CI analysis and production in support of DOD requirements. It focuses on hostile threat and foreign intelligence and security services, to include the development, population and maintenance of CI data bases for personalities, organizations, and installations (PO&I). PO&I files become the cornerstone of the CI activities planning and targeting, and guide CI and HUMINT activities. DIA's Defense HUMINT Service (DHS) is the force provider for strategic HUMINT forces and capabilities. During operations, elements from DHS form a partnership within the supported JTF headquarters' J-2X element for the coordination and deconfliction of HUMINT-source related collection activities.

CHAPTER 4. COUNTERINTELLIGENCE EMPLOYMENT

4001. OPERATIONAL ENVIRONMENT

As forces are committed to an operation, the threat picture expands and situational awareness improves. As U.S. military involvement increases, existing threats remain and may increase while new threats may emerge. In humanitarian operations and other MOOTW, the principal threats facing the MAGTF are criminal, terrorist, and espionage. These threats continue into the upper levels of conflict, with the addition of threats posed by irregular forces, special operations forces, and finally, by large-scale conventional military forces.

Within MOOTW and in lesser levels of conflict where there may be no designated MAGTF or joint rear area, CI activities are directed at supporting force protection efforts by engaging with key civic leaders, existing intelligence and security structure, factional leaders and cooperative personnel, and allied forces. Threats are normally at the low to mid level. Threats at higher levels of conflict normally involve conventional or unconventional force threats that require combat forces to counter. MAGTF CI elements conduct actions in support of these operations within the MAGTF area of operations and other assigned sectors as directed (e.g., the joint rear area). Three basic political operational categories can be used to frame CI activities. These are—

The primary operational environmental factor influencing MAGTF CI activities is political, vice physical.

- **Permissive**—An operational environment that specific agreements allow CI to conduct activities independently or with the host nation. In these environments, MAGTF CI activities and support to the security posture of the deployed forces are normally conducted with the host nation, or the host nation has provided concurrence, either direct or tacit.
- **Semi-Permissive**—An operational environment where there are either no in-place government organizations and/or laws, or where the government in power is not duly recognized by the U.S. or other international bodies. In these situations, the rules of engagement established by the JTF or multinational force commander is often the key variable as the host nation's civil, military, and security agencies are frequently degraded or nonexistent (or may even be supporting threat forces). Rules of engagement of the deploying force primarily drive limitations and restrictions that may be placed on CI activities.
- **Non-Permissive**—A non-permissive operational environment is one in which U.S. CI activities and contacts with the host nation are extremely limited, normally at the direction of the host nation. The situation in these countries may also place the U.S. in a situation where the actions of the host nation or individuals in the host nation government may be inimical to those of the U.S. government. In most cases, the host government may severely curtail contacts, normally only through a single point of contact. In some cases, the information provided may be of questionable validity.

MAGTF CI personnel must be aware of the differences in each operational environment and be able to establish operations based on the mission, nature of the environment and threat conditions.

Since much of the threat and vulnerability intelligence will be based on information provided by host nation or other sources, vulnerability assessments must be assessed and rapidly updated as necessary.

CI activities help the MAGTF commander develop estimates of the situation, shape the battlespace, and guide intelligence and force protection operations. The MAGTF commander focuses the CI effort by clearly identifying priority intelligence requirements, careful assignment of CI missions and tasks, and a clear statement of desired results. By orienting CI capabilities, the commander guides who or what are the CI targets and the nature and focus of operations (e.g., whether CI efforts will be designated primarily as defensive CI operations or offensive HUMINT operations).

4002. EMPLOYMENT OF CI ELEMENTS

Command and Control and Concept of Operations

The METT-T factors and the tactical concept of operations govern C2 relationships and the execution of CI plans and operations within the MAGTF. Specific CI concept of operations and C2 relationships will be established in either annex B (intelligence) to the operations order, fragmentary orders or other directives.

General Service

CI elements normally operate in GS of the MAGTF. Operational control of CI elements by the force commander provides the commander with the means to meet the specific operational requirements of the MAGTF and other supported forces with limited organic CI resources.

Direct Support and Attachment

Situational and operational factors may require some CI elements to be either attached or placed in direct support of MAGTF subordinate elements (e.g., during operations involving widely separated units in areas of dense population). In such cases, supported unit commanders employ CI personnel to satisfy their CI requirements or other mission specified by the MAGTF commander.

Technical Control

Regardless of the type C2 relationships established, the MAGTF commander will retain technical control (TECHCON) authority over all MAGTF CI and supporting elements, which will be exercised via the G-2/S-2 and intel bn commander/ISC.

Concept of Employment

MAGTF CI elements can be deployed on an area coverage concept or by unit assignment.

Area Coverage

Geographic AOR. MAGTF CI elements employed under area coverage are assigned a specific geographic AOR. Under area coverage, CI support is provided to commands located within the designated area. CI elements continue to operate within the assigned area even though the tactical situation or supported units operating in the area may change. (See figure 4-1.)

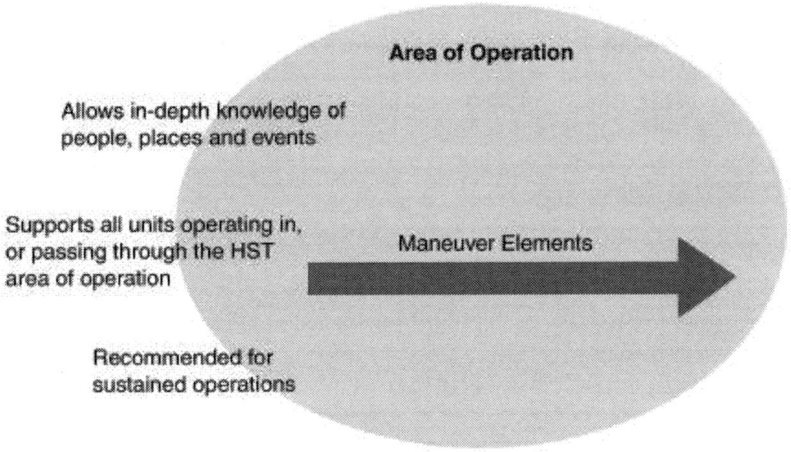

Figure 4-1. CI/HUMINT Support Using Area Coverage.

Continuity. Area coverage provides the greatest continuity of tactical CI operations. It allows MAGTF CI operations to focus on the enemy's intelligence organization and activities while remaining unfragmented and unrestricted by the tactical areas of responsibility assigned to support units. It also allows MAGTF CI personnel to become familiar with the area, enemy intelligence organization and operations, and CI targets. Area coverage is particularly effective during MOOTW (e.g., counterinsurgency operations) where the threat forces often operate on political, vice military, boundaries.

Unit Assignment

MAGTF CI elements employed on a unit assignment basis normally remain with designated supported units. They operate within that unit's AOR under the specified C2 relationships. As tactical units displace, it is necessary for higher echelons to provide CI coverage for the areas vacated. Relief of an area can be accomplished by three methods—attachment, leapfrog system, and relay system.

Attachment. CI elements may be detached from the MAGTF CE and attached to subordinate commanders. This method provides dedicated CI support under the operational control of the commanders to which the CI elements are attached for employment against specific CI targets during the initial (and possibly subsequent) phase of an operation. It also prepares for subsequent transfer of areas of responsibility from subordinate units to MAGTF CE without loss of continuity. These CI elements generally operate under the operational control of the supported commander during the initial reduction of CI targets. The CI elements then remain in place as the unit advances, reverting to the operational control of the MAGTF commander (or other specified commander) or intel bn commander. By remaining in place, the CI element ensures continuous coverage regardless of tactical unit movements.

Leapfrog System. This method is similar to the area coverage concept but on a smaller scale. Under the leapfrog system, MAGTF CI elements initially responsible for a specified area are detached from subordinate units and a new team attached as the operation progresses. The new CI element is attached sufficiently in advance to permit it to become thoroughly familiar with current operations within the AOR. This method of relief permits the CI element familiar with the area, informants, and targets to remain and conduct more extensive operations, while providing necessary CI direct support to subordinate commanders.

Relay System. This method requires MAGTF CI elements to be held in reserve. As the subordinate units advance, CI elements are dispatched forward to assume control of designated areas on a rotational basis.

CI Employment Considerations

Characteristics of the AOR influence the nature and extent of MAGTF CI operations. The following factors influence CI task organization, C2, and resulting concepts of operations and support relationships.

- Historical and recent espionage, sabotage, subversion or terrorism activities within the AO.
- Population density.
- Cultural make-up of the civilian population.
- Attitude of the people and political groups toward friendly and enemy forces.
- People's susceptibility to enemy penetration (hostile intelligence threat) and propaganda.
- Stability of the local government, security, and law enforcement.

The number of MAGTF CI resources available, particularly HSTs, is critical. Careful planning, awareness of CI operations throughout the joint AOR, and detailed intelligence and operations preparation are required. CI targets that require early reduction must be selected and the employment of MAGTF CI operations planned. Care must be taken to not overestimate CI element capabilities-this risks overextending and dispersing CI activity on many targets with limited effectiveness.

During amphibious operations, the commander, landing force (CLF) assumes operational control over assigned CI assets. Clear responsibility for CI operations must be assigned among MAGTF, naval, and other supporting CI elements. CI investigations of a GS nature, particularly in rear areas, may be tasked to NCIS or other supporting CI elements. In such cases, jurisdiction must be clearly defined to optimize overall CI support.

Employment of MAGTF CE CI Elements

MAGTF CE CI elements normally operate in the rear area conducting the following tasks:

- Provide overall MAGTF CI operational management and technical direction.
- Coordinate and integrate of MAGTF CI operations with JTF, multinational, and other supporting CI operations.

- Conduct CFSO and HUMINT operations of a relatively long-term nature.
- Provide assistance on military and civil security matters.
- Follow-up and complete CI tasks initiated by subordinate elements.

The MAGTF commander normally retains OPCON over technical surveillance countermeasures, CI inspections, and CI surveys for the entire MAGTF, which are exercised via the G-2/S-2 and the ISC. If required, intel bn may also establish and operate the MAGTF CI interrogation center.

Employment of CI Elements with the Ground Combat Element

Generally, the number of MAGTF CI elements employed in the ground combat element (GCE) AOR is greater than in other areas. During combat and in enemy occupied areas, the enemy has more opportunities to penetrate the CI screen because of the constant contact of opposing ground forces and the presence of indigenous or displaced populations. After the area has been cleared of the enemy, the CI element operating with the GCE is usually the first security unit to enter this area. They help determine initial requirements and establish initial security measures. CI elements with the GCE perform critical preparatory tasks for all subsequent CI and security operations. Prompt action by CI elements, particularly the rapid development and dissemination of intelligence based on interrogations or exploitation of captured materials and documents, can be of substantial benefit to the GCE's operations and force protection efforts.

The CI element focuses its operations on the GCE's distant and close areas, with responsibility for GCE rear area generally being coordinated with CI elements operating in GS of the MAGTF or those in direct support of the rear area operations commander. The CI elements operating with a division are generally deployed by task organizing multiple HSTs, although employment of CI teams is also an option. The HSTs and CI teams are responsible for the CI coverage of specific areas within the jurisdiction of the command. Each HST or CI team acts as an independent unit, but its activities are coordinated by the CI/HUMINT Company Commander or by one of the team OICs.

Time is essential during the initial phase of an operation. CI elements employed with attacking forces will generally limit their screening operation to identification and classification of enemy agents, collaborators, and civilians disguised as military personnel. If time permits, immediate tactical interrogation may be conducted of suspects. Normally, suspects will be passed to rear area, intermediate detention facilities, ultimately arriving at the Joint Interrogation and Debriefing Center for more detailed interrogations and classification.

CI elements must identify and secure—

- ·The most obvious CI targets.
- ·Agents left behind by the enemy for espionage and sabotage.
- ·Enemy collaborators.
- ·Key public buildings, such as the seat of the local government, police stations and communication centers.

Employment of CI with the Aviation Combat Element

There is no significant difference in the mission of CI elements employed with either ground or air units. However, air units are normally characterized by a static position. CI elements or personnel are attached or placed in GS of the tactical command echelon(s). They advise the commander on the control and security of sensitive areas, civilian control measures, and screening of local residents and transients. They also assist with the conduct of security assessments of facilities in the vicinity as required.

The aviation combat element (ACE) is often widely dispersed with elements operating from separate airfields. Since aircraft and support equipment are highly susceptible to damage and difficult to replace, aviation units are high priority targets for enemy saboteurs and terrorists. In many situations, ACE units employ large numbers of indigenous personnel in support roles, personnel who are a key target of enemy intelligence activities. Under such conditions, it may be necessary to provide HSTs in direct support of ACE elements.

CI Support to the Combat Service Support Element and Rear Area Operations

Within rear areas, the MAGTF CSSE is generally the principal organization requiring CI support. While the combat service support element's (CSSE) CI requirements are primarily concerned with military security, those of civil affairs elements generally deal with civil/military interaction and security of the populace. Despite the apparent differences of interest between their requirements, their CI problems are interrelated (e.g., a dissident civilian population hampering the efforts of MAGTF civil affairs elements attempting to establish effective administrative control in the area also disrupting logistics operations through sabotage, terrorism, and harassment attacks).

MAGTF CE CI elements performing GS to CI operations normally provide CI support to CSSE and other rear area operation elements. This includes support for installations and facilities dispersed through the combat service support areas. The number of team personnel supporting civil affairs units depends on the number of refugees to be identified in the area.

4003. FRIENDLY PRISONERS OF WAR AND PERSONS MISSING (NON-HOSTILE) AND MISSING IN ACTION

Friendly personnel who are captured by the enemy can be a source of information through the compromise of documents, personal papers or as the result of effective interrogation or coercion. It is a fundamental command responsibility to take necessary steps to counter any possible disclosure that would affect the immediate tactical situation.

CI units are assigned responsibility for investigating and determining risks posed to MAGTF operations by friendly personnel who have or may have been captured. They will help collect information of potential intelligence value on friendly personnel who may be under enemy control. They also collect intelligence information to aid in identifying, locating, and recovering captured friendly personnel. In addition, CI personnel conduct the intelligence and CI debriefings of friendly personnel who had been captured and then returned to friendly control.

When friendly personnel have or may have been captured, the identifying commander will immediately notify their unit intelligence officer, who will then coordinate with the pertinent CI element to initiate the CI investigative process (see appendix D). The CI unit may be able to immediately provide information that could aid in the search and recovery efforts, such as routes to enemy detention centers, locations of possible holding areas, and enemy procedures for handling and evacuating prisoners. If appropriate, the CI element can also initiate immediate CI collection action, such as using CI sources to gain information for possible recovery or search and rescue operations.

If the search or recovery attempts are unsuccessful, the CI unit initiates an immediate investigation to gather basic identification data and determine the circumstances surrounding the incident.

The investigation must be as thorough and detailed as possible and classified according to content. Every attempt is made to obtain recent photographs and handwriting samples of the captured person. A synopsis of the investigation, including a summary of the circumstances, is prepared on the CI report form. The completed basic identifying data form is attached as an enclosure to this report.

The investigation is designed to—

- Provide information to aid in subsequently identifying and locating the individual.
- Assess the potential intelligence value to the enemy.
- Collect intelligence information that will be of value when evaluating future intelligence reports.

In the case of aircraft incidents, the investigation includes type of aircraft, location, and sensitivity of classified equipment, bureau or registration number, call signs, and any aircraft distinguishing marks, such as insignia, etc. When feasible, the investigator should coordinate with the accident investigation team or aviation safety officer of the unit that experienced the loss.

The CI report, with the attached personnel data form, is distributed to the following commands:

- MARFOR component HQ, the MAGTF CE, and intel bn/detachment.
- Individual's parent command (division, Marine Aircraft Group or Force Service Support Group).
- Each CI element in the AOR.
- Other appropriate headquarters (e.g., combatant commander, Marine Corps Forces (MARFOR) headquarters, etc.).
- CMC (Counterintelligence Branch).

These reports are designed to aid follow-on CI operations. They do not replace normal G-1/S-1 casualty reporting procedures. When the CI report concerns a member of another Service assigned to a Marine Corps unit, a copy of the report is also provided to the appropriate component commander and Service headquarters. Subsequent pertinent information is distributed in the same manner as the initial CI report.

CI personnel debrief personnel returned to friendly control after being detained by the enemy. Normally, CI personnel supporting the unit that first gains custody of the individual conduct an initial debriefing to identify information of immediate tactical value and the locations of other friendly prisoners of war. As soon as possible, the returnee is evacuated to the MAGTF CE for further debriefing and subsequent evacuation to a formal debriefing site.

4004. UNIQUE CI SUPPORT DURING MOOTW

The CI operations previously discussed are generally applicable across the spectrum of MOOTW, including non-combatant evacuation, peace, and humanitarian and disaster relief operations. CI activities during MOOTW require CI personnel to be thoroughly familiar with the nature of operation, including its causes, characteristics, peculiarities, and with the threat infrastructure directing and controlling enemy efforts. Basic CI tasks are the denial of information to the threat force and the identification and neutralization of intelligence operations. A key aim of MOOTW is to restore internal security in the AOR, which requires a vigorous and highly coordinated CI effort.

Jurisdiction

The nature of the operation and the threat's covert methods of operation require the employment of a greater number of CI personnel than is generally required for conventional operations.

Effective CI operations require extensive coordination with the host country intelligence, CI, security, and law enforcement agencies. Operations are normally covered by a status of forces agreement (SOFA). A SOFA may include limitations and restrictions concerning the investigation and apprehension of host country citizens or other operations matters.

MAGTF CI Employment

Employment of CI elements during MOOTW is similar to that previously described. CI area coverage generally provides continuity of operations. As the threat's intelligence operations usually are well established, area coverage allows MAGTF CI personnel to better understand and more effectively counter the threat.

When assigning AORs, MAGTF CI elements ensure coverage overlaps to preclude gaps occurring between areas. Threat forces usually prefer to operate on the political or military boundaries where the assigned responsibilities of U.S. and allied forces may be vague and coordination is more difficult. CI elements employed through unit assignment are assigned responsibility for the area of interest around specified unit's area of responsibility. Under the unit assignment concept, rear area CI elements assume responsibility for any gap in coverage that may develop.

CI Measures and Operations

MOOTW usually require both passive and active CI measures be increased and aggressively pursued to effectively counter the threat's advantages and capabilities.

MAGTF units must institute and continuously enforce CI and security measures to deny information to the threat force and to protect friendly units from sabotage, espionage, subversion, and terrorism. In coordination with host country authorities, emphasis is needed on security measures and checks of indigenous employees or other persons with access to MAGTF installations, facilities, and command posts.

A significant factor during MOOTW is population and resources control. The movement channels and patterns necessary for support, communications, and operations of insurgent forces are observed and controlled. Prior to implementing control measures, the civilian population should be informed of the reasons for the controls. Whenever possible, such controls should be performed and enforced by host country agencies.

Basic CI operations, techniques, and procedures are generally applicable during MOOTW.

MAGTF CI elements must implement imaginative and highly aggressive special CI operations and HUMINT collection programs targeted against the threat's intelligence infrastructure and operational forces. The primary objective of special operations is the identification, location, and neutralization of specific members of the threat's infrastructure. A CI special operation consists of systematic intelligence collection and analysis with complete documentation concerning the activities of each targeted individual. This provides the host country with an account of the individual's illegal activities once the person is apprehended. Penetration of the infrastructure must be achieved at all levels possible—HUMINT operations are implemented to cover critical areas and to identify and locate threat forces. Intelligence derived from HUMINT programs may also be useful in CI special operations.

Cordon and search operations basically consist of—

- Security forces that surround the area, usually at night, to prevent persons from leaving the area.
- A sweep element that escorts detained people to a collection point at first opportunity.
- Search elements that conduct detailed en-mass searches of the area.
- Screening elements to process and screen detainees for identification of known or suspected threat personnel.

In MOOTW, cordon and search operations may be employed to ferret out the threat infrastructure. Ideally, a cadre of MAGTF CI personnel are assigned to each unit conducting the cordon and search operation to provide on scene exploitation and immediate reporting of threat related time-sensitive intelligence. These operations may also be employed to ferret out individual threat units that may use a community or area as cover for their activities or as a support base. Cordon and search operations should be conducted with host country forces and organizations, with U.S. forces including CI units providing support, advice, and assistance for the operation. At a minimum, host country personnel should be part of the screening and sweep elements of any cordon and search operation. Sweep/screening is often conducted with medical, civil affairs, and psychological operations programs that are accomplished after the screening phase. Throughout the operation, care must be exercised to prevent an adverse psychological effect on the populace.

Chapter 5. C2 and CIS Support to MAGTF CI Operations

5001. GENERAL

See MCWP 6-22, *Communications and Information Systems*, for a detailed review of MAGTF CIS doctrine and supporting tactics, techniques, and procedures.

The MAGTF CI effort depends heavily on secure, reliable, and fast communications and information systems (CIS) support to receive JTF, other components, theater, and national CI and all-source intelligence and to transmit organically collected and produced CI product and reports. CIS are also required for the command and control of MAGTF and supporting CI units and their integration with other intelligence and reconnaissance operations. Every mission and situation is unique, requiring some modifications to the supporting CIS architecture to support MAGTF CI operations. Detailed planning and close coordination between the CI/HUMINT company/detachment CO/OICs, the MAGTF G-2/S-2 and G-2/S-6, and pertinent operational and intelligence organizations is critical for establishing a reliable and effective CI CIS support.

5002. COMMAND AND CONTROL

JTF J-2 and the Joint Intelligence Support Element

General

The JTF J-2 organizational structure and capabilities will be situation and mission dependent as determined by the JFC and the JTF J-2. The JISE is the principal intelligence C2 node within the JTF J-2. The JISE is the focus for JTF intelligence operations, providing the JFC and component commanders with situational awareness and other intelligence support regarding adversary air, space, ground, and maritime capabilities and activities. CI collection, production and dissemination activities will be conducted within the JISE, or within the J-2's joint force J-2 CI/HUMINT staff element (J-2X), if established. Once initial basic and current CI products and support have been provided to the JTF and its components, updates will be accomplished by the JISE using push/pull dissemination techniques. Intelligence CIS based on the JDISS/Joint Worldwide Intelligence Communications System (JWICS) functionality provide the JTF with the ability to query theater and national CI organizations' servers and data bases for the most current CI support.

J-2X

Joint force commanders have operational control of JTF CI elements not organic to its component commanders, which they exercise via the JTF intelligence officer (J-2). Within the J-2, CI activities fall under the functional control of the TFCICA and HOC, which comprise the intelligence section's J-2X. CI collection operations management, CI production and CI dissemination tasks are exercised by the TFCICA, to include source deconfliction with the Central Source Registry, reporting coordination and resource application. Other functional managers, such as the Defense HUMINT Service representative within the J-2X or the HOC, have direct tasking authority over their functional assets, requiring close coordination and planning to ensure effective JTF CI/HUMINT operations.

National Intelligence Support Team

All-source national intelligence level CI and other intelligence assets may deploy in support of JTF (and even directly in support of MAGTF) operations to provide critical support via reach back and collaborative intelligence capabilities. The national intelligence support team (NIST) is the most typical method used. Its mission is to provide a tailored, national level all-source intelligence team to deployed commanders (generally at the JTF headquarters level, but support could be provided to other commands) during crisis or contingency operations. Depending on the supported unit's requirements, a NIST can task-organize to provide coordination with national intelligence agencies, analytical expertise, I&W, special assessments, targeting support, streamlined and rapid access to national intelligence data bases and other products, and assistance facilitating RFI management (see figure 5-1).

The NIST is a task-organized unit generally consisting of DIA, National Security Agency, Central Intelligence Agency, and, as appropriate, National Imagery and Mapping Agency personnel and equipment.

DIA, through the joint staff J-2, controls the NIST for deployment and administrative purposes. The composition and capabilities of each NIST deployment are unique based on the mission, duration, agencies representation, and capabilities required (see figure 5-2). During operations a NIST will usually be in direct support of the JFC, who exercises C2 via the JTF J-2. If a NIST is provides support of the JTF HQ, it generally will integrate its operations within the JISE. Key JISE functions and capabilities include collection management support, order of battle (OOB) analysis, identification of threat centers of gravity and critical vulnerabilities, and intelligence support to targeting and force protection.

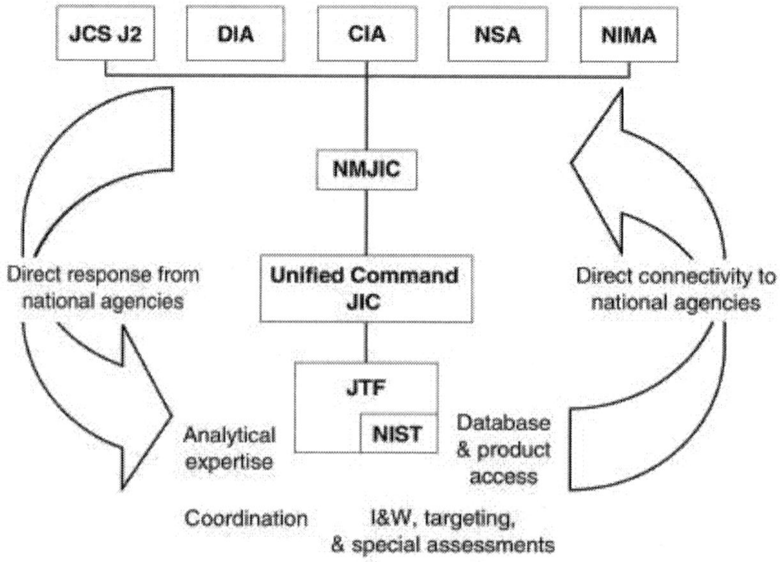

Figure 5-1. National Intelligence Support Team Capabilities.

A NIST is task-organized to fulfill the supported commander's intelligence requirements

Figure 5-2. Notional Composition of a National Intelligence Support Team.

Once deployed, any of the intelligence agencies with representatives on the NIST can provide its leadership. The basic C2 relationship between the NIST and the JTF (or other supported commands) is direct support. The NIST will be under the staff cognizance of the JTF J-2, performing intelligence support functions as so designated. The basic NIST concept of operations is to take the J-2's RFIs and collection and production requirements, discuss and deconflict these internally within the NIST to determine which element(s) should take these for action. Each NIST element leader, coordinated by the NIST team chief, will conduct liaison with their parent national intelligence organization. Intelligence and CI generated by the NIST will be disseminated to the JTF J-2 JISE or J-2X, the JFC, and other components of the JTF with the usual restriction based on clearance and programs.

A NIST's organic capabilities generally encompass only intelligence and some unique CIS support. NIST CIS capabilities will be task-organized. It may range from a single agency element's voice connectivity to a fully equipped NIST with joint deployable intelligence support system (JDISS) and Joint Worldwide Intelligence Communications System (JWICS) video teleconferencing capabilities (see figure 5-3 on page 5-4 for one of a NIST's key sophisticated CIS capabilities). Current methods of operation continue to rely on both agency and supported command-provided communications paths to support deployed NIST elements. The systems that elements are capable of deploying are discussed in greater detail in appendix C, NIST Systems, of JP 2-02, *National Intelligence Support to Joint Operations*.

Figure 5-3. NIST JWICS Mobile Integrated Communications System.

Amphibious Task Force Intel Center

During amphibious operations, amphibious task force (ATF) and the MEF CE's intelligence sections generally will integrate their operations. The principal intelligence C2 node is the amphibious task force intel center (ATFIC) located aboard the ATF flagship. The ATFIC is composed of designated shipboard spaces with installed CIS that support the intelligence operations of both the ATF and landing force while reducing duplicative functions and producing more comprehensive and timely intelligence support for the naval task force. Standard CIS connectivity is available—JWICS, SECRET Internet protocol router network (SIPRNET), nonsecure Internet protocol router network (NIPRNET), AUTODIN, DSN. Access is provided via the flagship's GENSER communication center and the special intelligence communications center within the ATFIC's ship's signals exploitation space.

MEF Command Element Intelligence C2 and Operations Nodes

Combat Intelligence Center and Intelligence Operations Center

The CIC and its subordinate elements is the principal MAGTF intelligence C2 node that provides the facilities and infrastructure for the centralized direction for the MEF's comprehensive intelligence, CI and reconnaissance operations. Since the CIC must effectively support the MEF, it must remain responsive to the requirements of all elements of the MAGTF. In supporting this objective, the CIC integrates and supports both MEF G-2 section and intel bn operations. While integrated, the organizational approach differs some for each of these.

CIC—overarching intelligence operations center established within the MEF main command post. Encompasses the primary functions of the MEF intelligence section and intel bn. It includes the sub-elements listed below.

G-2 Plans—main element of the G-2 section for coordinating and providing intelligence support to the MEF CE future plans team; and leadership and direction of the G-2 section's imagery and mapping, SIGINT, and weather sections.

G-2 Operations—main element of the G-2 section for coordinating and providing intelligence support to the MEF CE CG, battle staff and current operations center elements; target intelligence support to the force fires and future operations; G-2 section intelligence requirements management activities; Red Cell support; and MEF intelligence liaison with external commands and organizations.

IOC—principal MEF intelligence operations and C2 center that is established by intel bn. Performs intelligence requirement management, staff cognizance of ongoing organic and supporting collection operations, intelligence analysis and production, and intelligence dissemination.

- **Support Cell**—primary element for conducting MEF-wide intelligence requirements management; weather support; collections and dissemination planning and direction; and intelligence staff cognizance of MEF organic and supporting intelligence and reconnaissance operations.
- **P&A Cell**—primary analysis and production element of the MEF. Processes and produces all-source intelligence products in response to requirements of the MEF. It is the principal IMINT and GEOINT production element of the MEF.
- **Surveillance and Reconnaissance Cell (SARC)**—primary element for the supervision of MEF collection operations. Directs, coordinates, and monitors intelligence collection operations conducted by organic, attached, and direct support collection assets.
- **CI/HUMINT Company Command Post**—primary element for conducting CI/HUMINT planning and direction, command and control, and coordination of MEF CI/HUMINT operations with external CI/HUMINT organizations.

Operations Control and Analysis Center (OCAC)—main node for the C2 of radio battalion SIGINT operations and the overall coordination of MEF SIGINT operations. Processes, analyzes, produces, and disseminates SIGINT-derived information and directs the ground-based electronic warfare activities of the radio battalion.

Reconnaissance Operations Center (ROC)—main node for the C2 of force reconnaissance company's operations and the overall coordination of MEF ground reconnaissance operations. Processes, analyzes, produces, and disseminates ground reconnaissance-derived information in support of MEF intelligence requirements.

G-2 Section

The key G-2 nodes are organized to effectively align and support the MEF CE's staff cross-functional cellular staff organization and concept of operations. The G-2 plans branch provides intelligence and CI support to the MEF CE's future plans cell efforts. The G-2 operations branch, provides intelligence and CI support to the MEF CE's COC, FOC, force fires center and directs and manages the G-2's Red Cell and the MEF's external intelligence liaison teams (see figure 5-4).

CIC facilities, CIS, and other support must allow the AC/S G-2 and G-2 section to perform the following major tasks:

- Develop and answer outstanding MEF and subordinate units' PIRs and IRs by planning, directing, integrating, and supervising MEF organic and supporting intelligence, CI and reconnaissance operations.
- Plan the MEF concept of intelligence operations, including a concept for CI operations, for approval by the AC/S G-2 and subsequent implementation by the ISC based upon the mission, threat, commander's intent, guidance, and concept of operations.
- Recommend CI and force protection measures and countermeasures.
- Prepare appropriate intelligence and CI plans and orders for the MEF, including reviewing, coordinating, and integrating the intelligence plans of JTFs, theaters, and other organizations.
- Coordinate, provide, and facilitate the use of intelligence and CI to the MEF CG, the battlestaff, the future plans cells, the FOC, the COC, and the force fires center.
- Plan, direct, and supervise MEF liaison teams to external commands (e.g., the JTF and joint functional components headquarters) and intelligence organizations.
- Coordinate and supervise the transition of intelligence and CI planning and operations from G-2 plans to G-2 future operations, and from G-2 future operations to G-2 current operations, to effectively support the MEF's single battle transition process.

Intelligence Operations Center

The key subordinate elements within the IOC and their typical composition are the support cell, the SARC, and the P&A cell (see figure 3-3 on page 3-6).

The IOC is the other principal MEF CE intelligence node. It provides the facilities, CIS, and other support to allow the ISC and intel bn to perform the following tasks:

- Provide centralized direction for MEF intelligence and CI operations under the staff cognizance of the AC/S G-2. The IOC is the core for this task, with key assistance from the G-2 plans and G-2 operations elements.
- Consistent with the commander's priorities, consolidate, validate, and prioritize IRs of the entire force. The key CIC element providing for this is the collection management and dissemination (CMD) section within

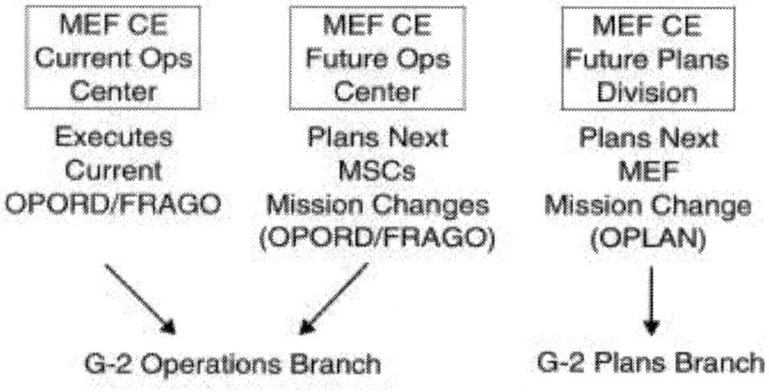

Figure 5-4. MEF CE Cross-Functional Cellular Organization and Intelligence Support.

the IOC's support cell. Intelligence specialists from all disciplines, including CI, generally are organic to this section.

- Plan, develop, and direct the MEF collection, production, and dissemination plans and operations. The key CIC elements providing for this are the CMD section within the IOC's support cell and the P&A cell.
- Submit consolidated requests for external intelligence and CI support through the Marine component headquarters to appropriate agencies. The key CIC element providing for this is the CMD section within the IOC's support cell, with assistance from the P&A cell and the G-2 operations branch.
- Allow the ISC to exercise, per AC/S G-2 cognizance, principal staff cognizance of MEF organic and supporting intelligence and reconnaissance operations, including CI, HUMINT, SIGINT, IMINT, GEOINT, measurement and signature intelligence (MASINT), ground reconnaissance, and aerial reconnaissance operations.
- Coordinate and manage the employment of MEF organic collection assets through the IOC's SARC. Within the SARC will be representatives from most organic and supporting intelligence, CI and reconnaissance units to provide C2 and reporting of ongoing intelligence operations.
- Maintain a consolidated, all-source intelligence production center in the MEF in the IOC's P&A cell. Other nodes with significant intelligence production involvement are the radio battalion's operations control and analysis center (OCAC) and the CI/HUMINT Co's CP. Similar to the CMD section, intelligence specialists from intelligence disciplines generally are organic to the P&A cell.
- Link the MEF CE to national, theater, joint, other-Service, and multinational intelligence and CI assets and operations. Intel bn, G-2 section C2, and operations nodes have common and unique capabilities to perform critical tasks within the function. In addition to MEF CE common communications pathways provided by the communications battalion, the IOC generally will also have unique intelligence communications capability, such as Trojan Spirit II.

Overall MEF Intelligence C2 Relationships

The MEF G-2 section and intel bn's overall command and control relationships and resulting all-source intelligence support flow throughout the MEF are as indicated in figure 5-5 on page 5-8.

CIC/IOC Operations and MAGTF CI Operations

Key CI activities, which will be integral to many CIC/IOC operations, include—

- **Collection.**
  - The CMD section, HQ, Intel Bn, provides the core for MEF CIC collection operations. During operations the CMD section is located within the IOC's support cell. Intelligence specialists from all disciplines, including CI, are organic to this section. Key CIS resources required include intelligence analysis system (IAS) and access to the full range of communications: (JWICS, SIPRNET, NIPRNET, DSN, etc.).
  - The SARC, another key element within the IOC, provides the other key component of collection operations. Within the SARC will be representatives from most organic and supporting intelligence

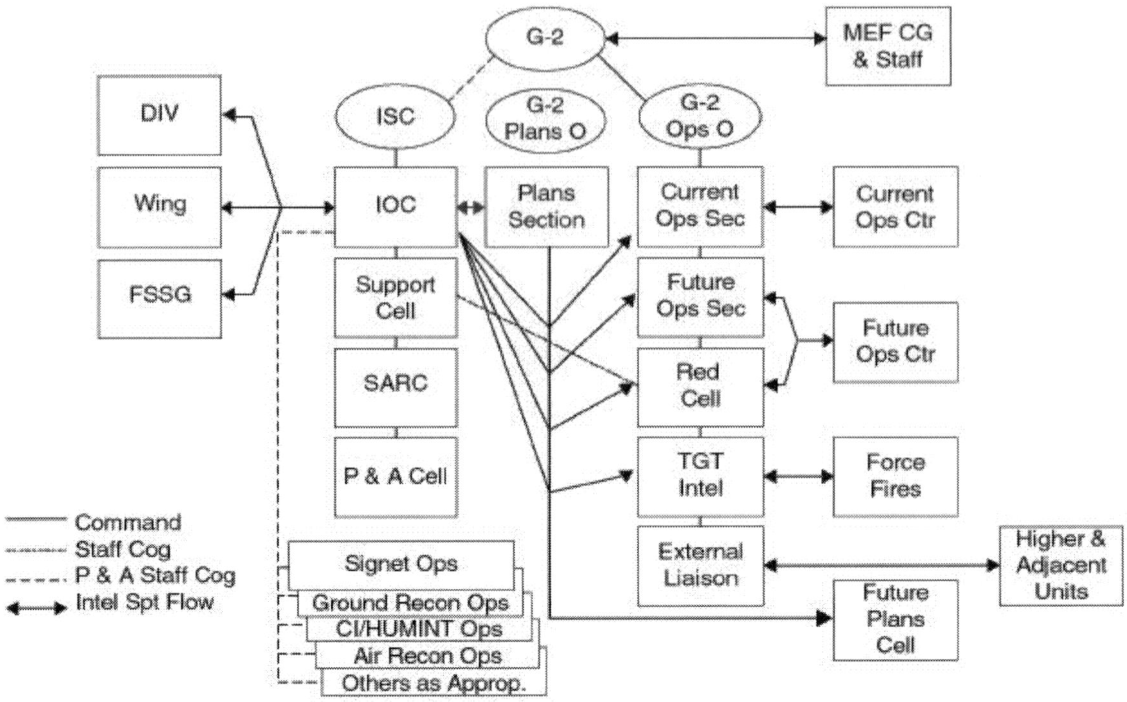

Figure 5-5. MEF G-2 and Intelligence Battalion C2 Relationships and MEF Intelligence and CI Support Flow.

and reconnaissance units providing C2 and reporting of ongoing intelligence operations. Regarding CI, the SARC will include representatives from CI/HUMINT Co to monitor ongoing CI/HUMINT operations and report time-sensitive intelligence.

- **Production.** The P&A cell, Intel Bn, provides the core for MEF intelligence production operations. Similar to collection, intelligence specialists from all intelligence disciplines are organic to the P&A cell. Key CIS resources required include IAS and JDISS, with access to the full range of communications (JWICS, SIPRNET, NIPRNET, DSN, etc.). Additionally, the operations/analysis element from CI/HUMINT company may be integrated into P&A cell operations to efficiently support both CI and all-source production operations.
- **Dissemination.** The CMD section, HQ, Intel Bn, provides the core C2 for MEF intelligence dissemination operations. Key CIS resources required include IAS and JDISS, with access to the full range of communications (JWICS, SIPRNET, NIPRNET, DSN, etc.) for external dissemination; and IAS via the TDN and other MEF communications resources for internal dissemination.

CI/HUMINT Company HQ

CI and IT platoons elements as well as task-organized HSTs generally will be employed throughout MAGTF. To support operations, the CI/HUMINT company headquarters will usually establish a command post (CP) in the vicinity of the IOC, integrated with and employing the full range of CIS supporting it.

CI/HUMINT Co C2, planning and direction, and some analysis, production, and dissemination functions are executed within the CI/HUMINT Co CP. The CI/HUMINT Co personnel within the CP perform the CI and HUMINT processing, analysis, exploitation, production, and reporting of CI and HUMINT products and information per ISC direction and the intelligence operations plan.

It is the principle element that coordinates with other intel bn and G-2 nodes to plan and direct CI/HUMINT operations.

Deployed CI/HUMINT Elements

CI/HUMINT elements—HSTs, CI teams, IT teams—will generally be deployed at many command echelons and locations within the MAGTF. For example, significant elements may be attached or placed in direct support of GCE forces or the rear area operations commander. Likewise, CI/HUMINT elements may be employed at the MAGTF EPW compound and other EPW collection points. The CIS resources used by these CI elements will be dependent upon the situation. Generally these elements will use company or intel bn CIS resources to satisfy their organic C2 needs, while using the supported unit's CIS resources for broader requirements.

5003. BASIC CI CIS REQUIREMENTS

Regardless of the size of MAGTF CI/HUMINT forces, there are certain standing CI CIS requirements must be satisfied. These requirements are—

- **Capability to command and control subordinate units.** Intelligence officers and CI/HUMINT element commanders/OICs must be capable of positive C2 of subordinate units and integration of its operations with broader MAGTF and external intelligence and operations C2. Traditionally single-channel radio and record message traffic have been used to support C2 of MAGTF CI units. In semi-static situations, secure E-mail or telephone may be the method of choice, while in highly fluid or mobile scenarios, cellular, SATCOM, and VHF and HF radio may be used.
- **Ability to receive collected data and information from deployed CI elements.** The CIS architecture must provide connectivity between organic and supporting CI/HUMINT elements (such as the HUMINT support teams or CI liaison elements), CI analysis and production centers, and supported MAGTF operations and intelligence centers. Requirements include the capability to transmit collection files and reports digitally via fiber optics, wire, or radio in formats (both voice and data) that are readily usable by the CI and all-source intelligence analysts.
- **Ability to provide intelligence to supported commanders.** CI CIS requirements will be influenced by supported commanders' intents, concepts of operations and intelligence, command relationships, and standing PIRs and IRs. The CIS architecture must be capable of integrating CI/HUMINT element C2 and supporting CIS operations (including special communications capabilities and channels unique to CI reporting) with the primary CIS channels used by supported commanders.
- **Ability to share CI products and reports with MAGTF all-source intelligence centers and with CI and all-source JTF, other components, theater, and national CI and intelligence centers.** The traditional means for providing this capability are MAGTF general service secure record and voice communications. While these techniques

continue to be used, they are rapidly becoming secondary in importance to the use of the JWICS, the SIPRNET, and CI unique CIS capabilities that allow participants to access each others CI products and data bases and immediately pull required data, intelligence, and CI products.

5004. CIS SUPPORT TO MAGTF CI OPERATIONS

General

CI CIS architecture for any given operation is dynamic. Key reference documentation with respect to a specific theater or MAGTF operation are—

- Combatant command, JTF, and MAGTF CI/HUMINT plans developed for various OPLANs.
- MAGTF command element intelligence SOP and combatant commanders intelligence and CI/HUMINT tactics, techniques, and procedures.
- Annexes B (intelligence), C (operations), J (command relationships), and K (communications and information systems) of the MAGTF and JTF OPORDs.
- The following parts of Annex B (Intelligence) to a MAGTF OPORD: Appendix 3 (Counterintelligence Operations); appendix 5 (Human Resources Intelligence Operations); and tab D (Intelligence Communications and Information Systems Plan) to appendix 16 (Intelligence Operations Plan).

Communications Systems

Information systems and supporting communications connectivity are evolving rapidly within the Marine Corps and other elements of the military. The following information provides typical key MAGTF CI communications requirements.

The MAGTF mission, the nature of the threat, friendly concepts of operations and intelligence, supporting task organization and command relationships, and extent of allied/ multinational operations are the key factors influencing what specific CI communications are established during operations.

Intelligence and CI/HUMINT Radio Nets

The following are radio nets typically established for either dedicated CI needs or are intelligence nets that CI elements may need to be stations on:

MAGTF Intelligence (UHF-SATCOM/HF/VHF). Used for rapid reporting and dissemination of intelligence, collaborative planning of future MAGTF intelligence operations, and C2 of ongoing MAGTF intelligence and reconnaissance operations. Typical organizations/elements participating in this net include: the MAGTF CE; the GCE/ACE/CSSE headquarters; intelligence, CI and reconnaissance elements either attached to, OPCON or supporting the MAGTF; and others as directed.

GCE/ACE/CSSE Intelligence (HF/VHF). Used to provide rapid reporting and dissemination of intelligence, collaborative planning of future intelligence operations, and command and control of ongoing intelligence and reconnaissance operations. Typical organizations/elements participating in this net include: the GCE/ACE/CSSE headquarters; headquarters of their major subordinate units; intelligence, CI and reconnaissance elements either attached to, OPCON or supporting the MSE headquarters; and others as directed.

CI/HUMINT Team(s) Command (HF/VHF). Used for C2 of CI teams, IT teams, and HSTs operations, and the coordination of CI/HUMINT administrative and logistic support. This net will also generally terminate in the MAGTF SARC.

CI/HUMINT Reporting Net (VHF/HF). Used as a means for the rapid reporting of CI/HUMINT data to supported units. Participants generally include CI teams, IT teams, and HSTs operations, the SARC, and the intelligence centers of any supported units.

CI/HUMINT Communications Equipment

CI elements require extensive communications support from the command to which they are attached. These requirements include secure dedicated and shared systems connectivity and are situationally dependent based upon employment method, terrain, distance, and other factors. CI elements usually deploy with the following organic communications and information systems:

- **SINCGARS radios**—primarily to support CI element C2 and intelligence reporting.
- **Motorola SABER radios**—principally to support internal CI element communications of an operational nature (e.g., surveillance or security).

The command provides frequency management, cryptographic materials system control, and logistics support (e.g., batteries and maintenance).

Intelligence and CI/HUMINT Information Systems

The following systems and data bases are established for either dedicated or multipurpose use.

Joint Deployable Intelligence Support System

JDISS is an all-source automated intelligence tool that provides the backbone of intelligence connectivity among the national, theater, JTF headquarters, component commanders, and other intelligence organizations. JDISS employs a transportable workstation and communications suite that electronically extends JIC capability to a JTF and other tactical users. SIPRNET or JWICS will provide the principal communications connectivity for JDISS.

Intelligence Analysis System

IAS provides automated applications and other tools for MAGTF all-source intelligence planning and direction, management, processing and exploitation, analysis and production, and dissemination. Various configurations of IAS will be organic to intelligence sections from the battalion/squadron through MEF CE levels.

Defense Counterintelligence Information System

Defense counterintelligence information system (DCIIS) is a DOD system that automates and standardizes CI functions at command echelons. DCIIS contains standardized DOD forms (for CI investigations, collections, operations, and analysis and production) and shared CI data bases. DCIIS also contains supplemental forms to satisfy tactical reporting requirements of Marine and Army CI elements. DCIIS is interoperable with JDISS and other intelligence systems.

Defense Intelligence Threat Data System

Defense intelligence threat data system (DITDS) is available via JDISS, IAS and other intelligence systems. It contains the DOD CI/counter-terrorism/counter-proliferation data bases and is principally used by CI analysts and production personnel. The system provides a number of analytical tools, such as automated and graphical link analysis hot-linked to the underlying reports, automatic time lining and access to various communications systems to support dissemination. Communications connectivity is via SIPRNET

Migration Defense Intelligence Threat Data System

Many of the current intelligence/CI data bases that reside on other systems will become resident within the MDITDS. Some of these include DITDS, SPHINX (DIA's CI data base), CANNON LIGHT (U.S. Army CI data base), BLOODHOUND (EUCOM CI data base), Automated Intelligence Information Retrieval System, Automated Decisionmaking and Program Timeline, Defense Automated Warning System, Sensitive Compartmented Automated Research Facility, Terrorism Research & Analysis Program (TRAP), and many others.

The migration defense intelligence threat data system (MDITDS) is being developed to operate on the DCIIS to provide an automated production system for DODIIS I&W, CI, counterterrorism, and Arms Proliferation/Defense Industry communities.

HUMINT Operational Communications Network

HUMINT operational communications network (HOCNET) is the umbrella name given to a collection of systems and applications currently operational or under development that support the Defense HUMINT Service (DHS) worldwide activities (i.e., Defense Attaché Worldwide Network).

Special Operations Debrief and Retrieval System

Special operations debrief and retrieval system (SODARS) provides detailed mission debriefs and after-action reports from special operations forces (SOF). As SOF are often the first and perhaps only DOD force committed to an operation, they may provide invaluable intelligence and CI when developing pre-deployment threat assessments.

AN/PYQ-3 CI/HUMINT Automated Tool Set

AN/PYQ-3 CI/HUMINT automated tool set (CHATS) consists of CIS hardware and software designed to meet the unique requirements of MAGTF CI/HUMINT elements. Operating up to the SECRET level and using the baseline DCIIS software suite, the system provides the capability to manage MAGTF CI assets and analyze information collected through CI investigations, interrogations, collection, and document exploitation. With CHATS, CI units may electronically store collected information in a local data base, associate information with digital photography, and transmit/receive information over existing military and civilian communications. (See appendix A for additional information on CHATS.)

Joint Collection Management Tool

Joint collection management tool (JCMT) is the principal automated tool for all-source intelligence collection requirements management. It provides a capability for management, evaluation, and direction of collection operations. Using DCIIS, CI collection requirements and taskings can be accessed from the JCMT.

Community On-Line Intelligence System for End-Users and Managers (COLISEUM) System

Community on-line intelligence system for end-users and managers (COLISEUM) is the automated, DOD intelligence production program requirements system that allows authorized users to directly submit multi-discipline intelligence production requirements to commanders and intelligence production centers. COLISEUM also tracks responses and provides status reports on validated production requirements.

Summary

Figure 5-6 on page 5-14 depicts a notional MEF CI architecture, and figure 5-7 on page 5-15 depicts key CI elements and supporting CIS within the MEF CE CIC and IOC. The three key aspects of MAGTF CI C2 and supporting CIS operations are—

- Task organization and command/support relationships of MAGTF CI units—CI/HUMINT Co headquarters collocated with the MAGTF G-2/S-2 CIC/IOC. Although company elements normally operate in GS of the MAGTF, task-organized CI, IT or HSTs may be either attached or placed in direct support of MAGTF subordinate units.
- Principal CI systems (e.g., CHATS) employed within and in support of the MAGTF.
- Communications connectivity—the principal communications pathways and the level of security classification.

5005. CI CIS PLANNING CONSIDERATIONS

The following identifies key CIS requirements and planning considerations supporting MAGTF CI operations.

- Ensure that the MAGTF CE G-2/S-2, intel bn, CI/HUMINT Co elements, and other MAGTF units are included in the distribution of CI/HUMINT-related address indicator groups to receive pertinent JTF, theater, and national intelligence and CI products.
- Determine and coordinate radio net requirements, supporting frequencies, and operational procedures supporting CI operations (external to MAGTF, internal MAGTF, intelligence broadcasts, retransmission sites, routine and time-sensitive operations, etc.).
- Coordinate CI CIS activation and restoration priorities and supporting procedures.
- Determine cryptograph material system requirements for unique CI communications.
- Determine and coordinate wire communications (including telephones) supporting CI operations.
- Establish, operate, and manage unique CI communications.
- Determine and coordinate local and wide area networks and unique intelligence networks information systems requirements in support of CI operations (hardware, software, Internet protocol addresses, etc.).

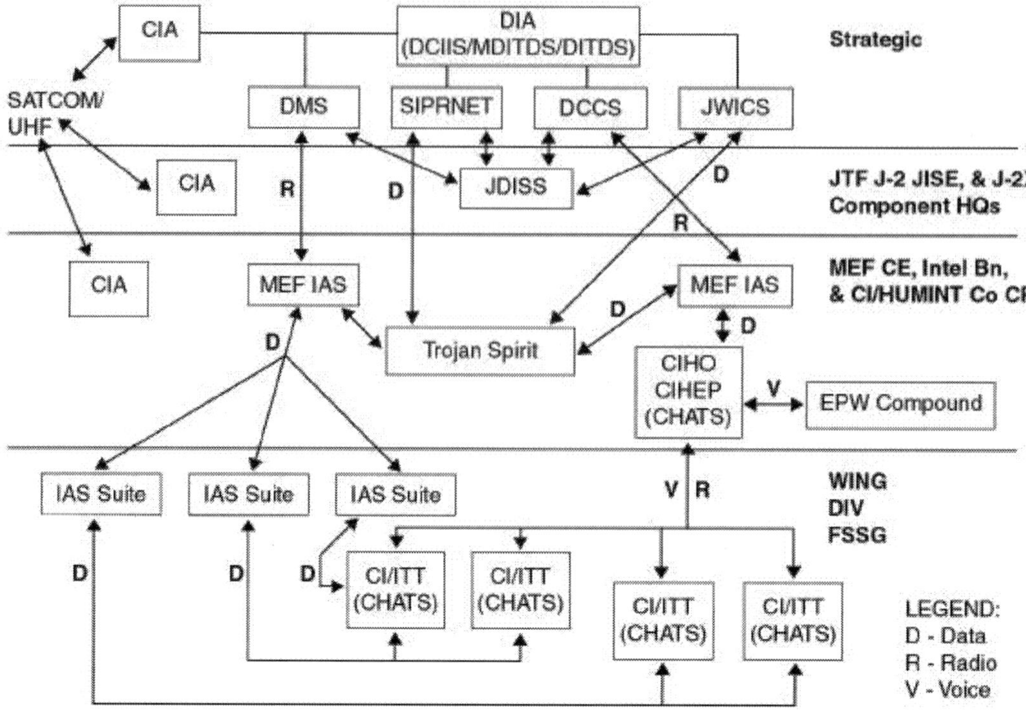

Figure 5-6. Counterintelligence Architecture.

- Integrate CI/HUMINT elements' CIS operations with those of other MAGTF and pertinent JTF and other components intelligence and reconnaissance units (mutual support, cueing, etc.).
- Integrate communications of CI/HUMINT elements employed in GS with collocated GCE, ACE, CSSE, and other MAGTF elements (e.g., to provide time-sensitive reporting, coordination of maneuver, etc.).
- Coordinate CI CIS and dissemination operations and procedures with allied and coalition forces.

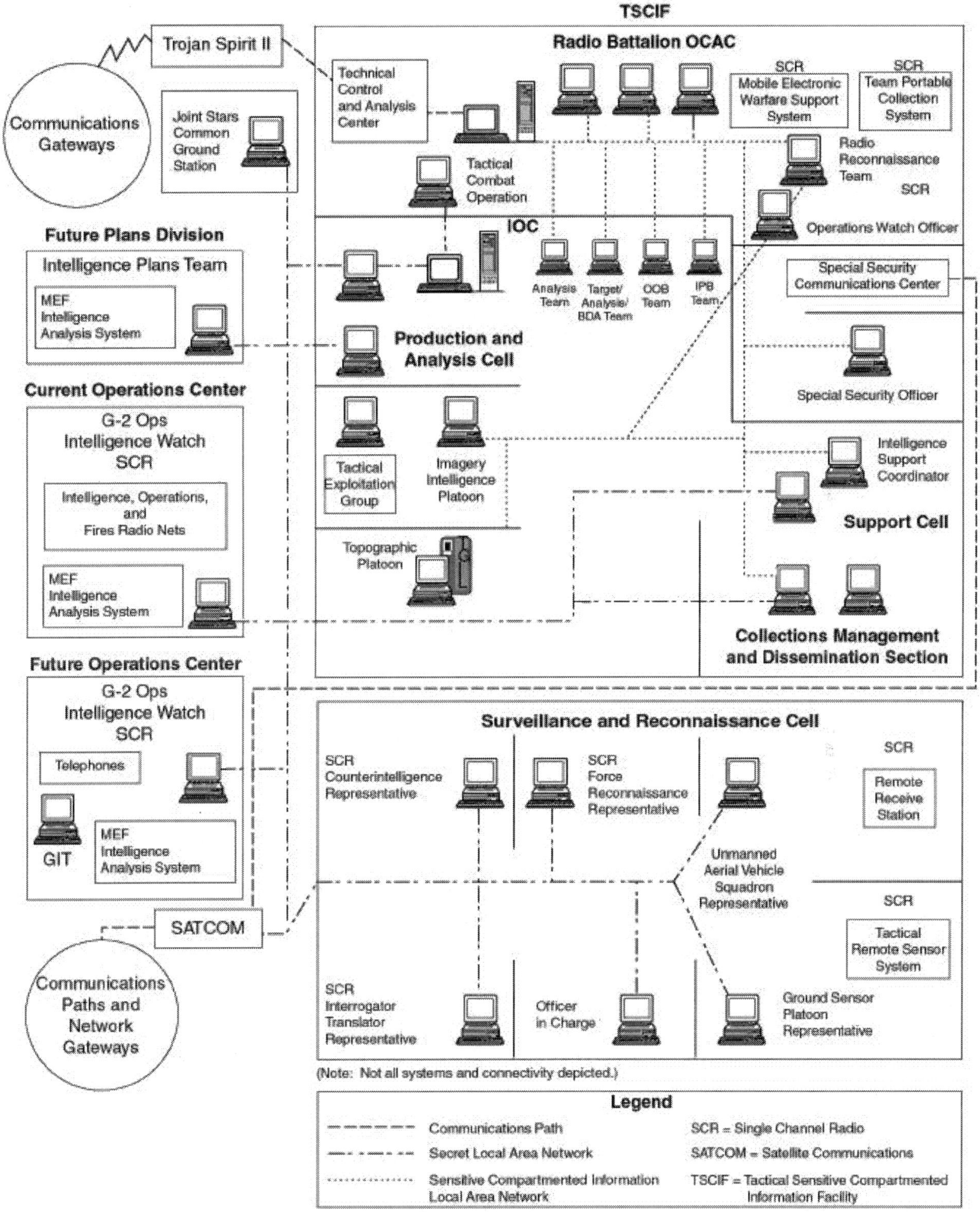

Figure 5-7. CI Elements within the MAGTF CE CIC and IOC and Key Communications and Informations Systems.

CHAPTER 6. COUNTERINTELLIGENCE PLANNING

6001. MARINE CORPS PLANNING PROCESS AND JOINT PLANNING PROCESS OVERVIEW

Planning is an act of preparing for future decisions in an uncertain and time-constrained environment. Whether it is done at the national or the battalion/squadron level, the key functions of planning are—

- Planning leads to a plan that directs and coordinates action.
- Planning develops a shared situational awareness.
- Planning generates expectations about how actions will evolve and affect the desired outcome.
- Planning supports the exercise of initiative.
- Planning shapes the thinking of planners.

Marine Corps Planning Process

See MCWP 5-1, *Marine Corps Planning Process*, for detailed doctrine and TTP regarding the MCPP. JP 5-00.2, *Joint Task Force Planning, Guidance and Procedures* provides detailed discussion of the joint planning process.

The Marine Corps Planning Process (MCPP) helps organize the thought processes of commanders and their staff throughout the planning and execution of military operations. It focuses on the threat and is based on the Marine Corps warfighting philosophy of maneuver warfare. It capitalizes on the principle of unity of effort and supports the establishment and maintenance of tempo. The MCPP steps can be as detailed or as abbreviated as time, staff resources, experience, and the situation permit. It applies to command and staff actions at all echelons. From the Marine Corps component headquarters to the battalion/squadron level, commanders and staff members must master the MCPP to be full participants in integrated planning. Additionally, the MCPP complements deliberate or crisis action planning (CAP) as outlined in the Joint Operation Planning and Execution System (JOPES).

The MCPP (see figure 6-1 on page 6-2) establishes procedures for analyzing a mission, developing and analyzing COAs against the threat, comparing friendly COAs against the commander's criteria and each other, selecting a COA, and preparing an OPORD for execution. The MCPP organizes the planning process into six steps. It provides commanders and their staff a means to organize their planning activities and transmit the plan to subordinates and subordinate commands. Through this process, MAGTF levels of command can begin the planning effort with a common understanding of the mission and commander's guidance.

The six integrated steps of this process are—

- **Mission analysis.** Mission analysis is the first step in planning. The purpose of mission analysis is to review and analyze orders, guidance, and other information provided by higher headquarters and produce a unit mission statement. Mission analysis drives the MCPP.
- **COA development.** During COA development, the planners use the mission statement, including higher headquarters tasking and intent,

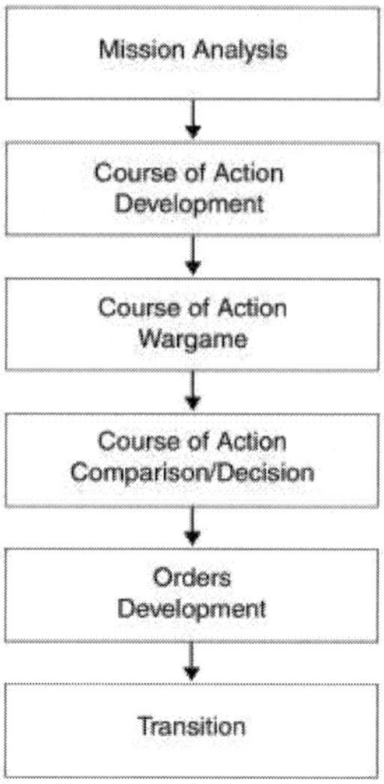

Figure 6-1. The Marine Corps Planning Process.

commander's intent, and commander's planning guidance to develop the COA(s). Each prospective COA is examined to ensure that it is suitable, feasible, distinguishable, acceptable, and complete with respect to the current and anticipated situation, the mission, and the commander's intent. Per the commander's guidance, approved COAs are further developed in greater detail.

- **COA wargaming.** During COA wargaming, each friendly COA is examined against selected threat COAs. COA wargaming involves a detailed assessment of each COA as it pertains to the threat and the environment. It assists the planners in identifying strengths and weaknesses, associated risks, and asset shortfalls for each friendly COA. It identifies branches and potential sequels that may require additional planning. Short of actually executing the COA, COA wargaming provides the most reliable basis for understanding and improving each COA.
- **COA comparison and decision.** In COA comparison and decision, the commander evaluates all friendly COAs—against established criteria, then against each other—and selects the COA that he deems most likely to accomplish the mission.
- **Orders development.** During orders development, the staff takes the commander's COA decision, mission statement, commander's intent, guidance, and develops orders to direct the actions of the unit. Orders serve as the principal means by which the commander's decision, intent, and guidance are expressed.
- **Transition.** Transition is an orderly handover of a plan or order passed to those tasked with execution of the operation. It provides those who will execute the plan or order with the situational awareness and rationale for key decisions necessary to ensure there is a coherent shift from planning to execution.

Comparison of the MCPP and the Joint Planning Processes

Joint Deliberate Planning

The deliberate planning process is used by the joint staff and commanders in chief (CINCs) to develop plans (OPLANs, CONPLANs, functional plans) supporting national strategy. The Joint Strategic Capabilities Plan apportions forces and resources for use during deliberate planning by the combatant commanders and their service component commanders. Figure 6-2 illustrates how the MCPP fits within and supports the joint deliberate planning process.

Interactions among various planning steps allow a concurrent, coordinated effort that maintains flexibility, makes efficient use of time available, and facilitates continuous information sharing.

Crisis Action Planning (CAP)

CAP is conducted in response to crises where national interests are threatened and a military response is being considered. In CAP, the time available for planning at the national level may be as little as a few days. CAP procedures promote the logical, rapid flow of information and the timely preparation of campaign plans or OPORDs. The figure 6-3 on page 6-4 illustrates how the MCPP fits within and supports the joint crisis action planning process.

6002. CI PLANNING

Intelligence Planning

CI planning and execution is conducted in concert with the six phases of the standard intelligence cycle. The first phase is planning and direction. It consists of those activities that identify pertinent intelligence requirements (IR) and provides the means for satisfying those requirements. Intelligence planning and direction is a continuous function and a command responsibility. The commander directs the intelligence effort; the intelligence officer manages this effort for the commander based on the intent, designation of priority intelligence requirements (PIRs) and EEFI, and specific guidance provided during the planning process.

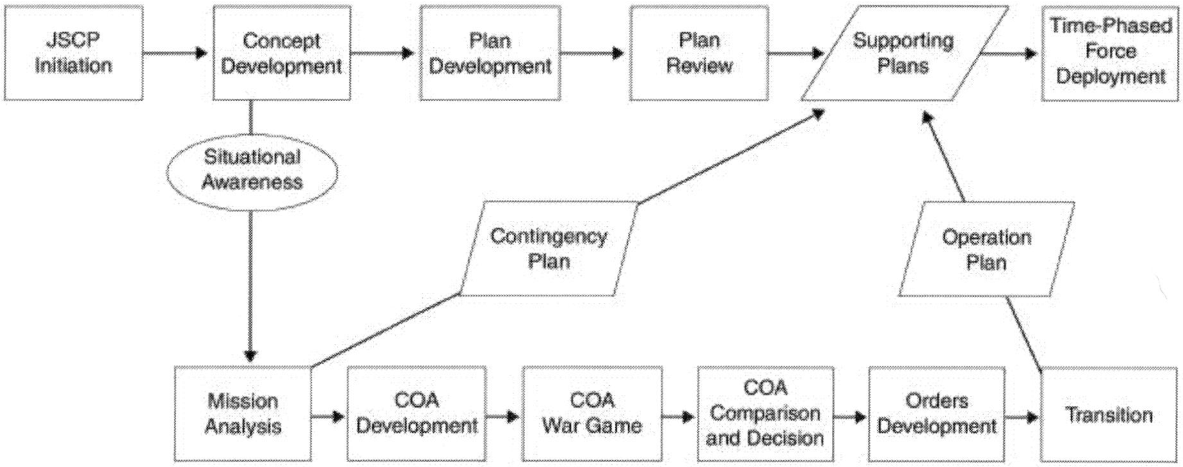

Figure 6-2. The MCPP and the Joint Deliberate Planning Process.

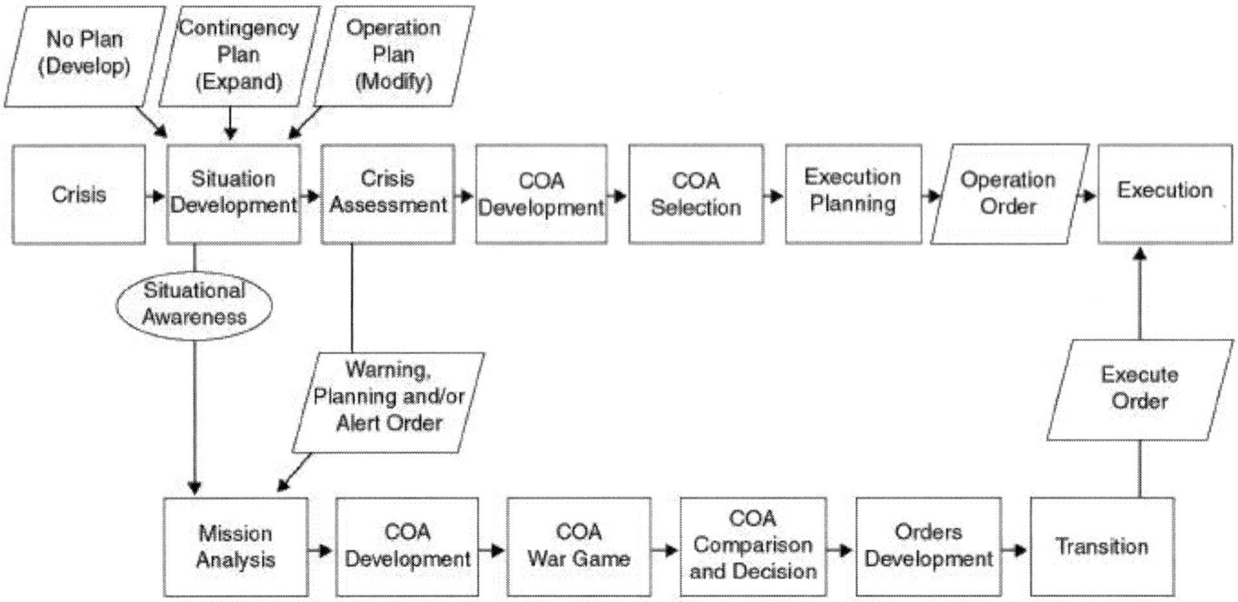

Figure 6-3. The MCPP and the Joint Crisis Action Planning Process.

CI Planning—General

Focus

CI planning and subsequent operations are conducted in support of the MAGTF or designated subordinate commanders to support the overall intelligence effort and to aid with force protection. Accordingly, CI must be planned with the overall intelligence and force protection efforts. The commander must incorporate CI early in the planning process to formulate an estimate of the situation, identify the MAGTF's risks and security vulnerabilities, and begin shaping overall and supporting intelligence and force protection operations.

Planning and Direction Functions

- Requirements development.
- Requirements management.
- Collection management.
- Production management.
- Dissemination management.

CI and HUMINT

CI must be considered with many other intelligence activities because of the mission of CI. HUMINT is the intelligence activity that has the most important connective tie to CI. CI and HUMINT work hand-in-hand because of the nature of their targets and the type of intelligence missions they perform: CI neutralizing the enemy intelligence effort and HUMINT collecting information about enemy activity. The differentiation and coordination required for the effective exploitation of human sources is critical, requiring integration of CI activities and HUMINT operations (e.g., IT exploitation of EPW). See DIAM 58-11, *DOD HUMINT Policies and Procedures*, and DIAM 58-12, *The DOD HUMINT Management System*, for additional information.

See chapter 3 of MCWP 2-1 for comprehensive discussion of each phase of the intelligence cycle, intelligence requirements management, and the overall conduct of intelligence planning and direction.

CI Planning Responsibilities

See paragraphs 3002 and 3003.

Coordination Considerations

The complexity of CI operations requires thorough coordination with intelligence organizations. Detailed coordination ensures that CI operations are focused on intelligence priorities and are not duplicative. Constant coordination must be accomplished with the JTF and other component forces, combatant commander, and other supporting intelligence organizations to ensure coordinated, manageable, and effective CI operations are conducted.

CI Planning Considerations

Key considerations in planning CI operations include—

Friendly Considerations. CI operations must support and adapt to the commander's intent, concepts of intelligence and operations, and the supporting scheme of maneuver. Questions to answer include—

- What are the MAGTF areas of operations (AO) and areas of interest (AI)?
- What are the MAGTF concept of operations, task organization, main and supporting efforts?
- What are the task organization and the C2 command/support relationships among MAGTF intelligence, CI and reconnaissance units? Can the friendly concept of operations be supported by CI elements operating in MAGTF GS, or are CI direct support/attachments to MAGTF subordinate units required? What are the standing PIRs and IRs? Which have been tasked to supporting CI units? What specific information is the commander most interested in (i.e., enemy air operations, enemy ground operations, friendly force protection, target BDA, or enemy future intentions)?
- What is the MAGTF force protection concept of operations? What are the standing EEFIs and their assessed priorities?
- What are the CI, intelligence and force protection concepts of operations of the JTF, other components and theater forces? How can external CI assets be best integrated and employed to support MAGTF operations?

Enemy Considerations

Intelligence operations focus on the enemy. Prior to commencing MAGTF CI operations, we must learn as much as we can about the enemy. Key adversary information which must be considered when conducting CI planning include—

- What threat forces—conventional, law enforcement/security, paramilitary, guerrilla, terrorist—are within the MAGTF AO and AI? What are their centers of gravity and critical vulnerabilities? Is this a large enemy force organized along conventional military lines or a small, loosely knit guerrilla or unconventional military force? What are their sizes, composition, tactics, techniques, and procedures?
- Who are the key enemy military, security, and civilian leaders? What and where are the enemy's critical nodes for C2 and what are their vulnerabilities? What security countermeasures do they employ to prevent CI exploitation of their operations? What are its C2 and CIS tactics, techniques, and procedures?

- Who are the known enemy personalities engaged in intelligence, CI, security, police, terrorist, or political activities? Who are known or suspected collaborators and sympathizers, both within the populace and within other parties?
- What are the key installations and facilities used by enemy intelligence, espionage, sabotage, subversive, and police organizations? What are the key communications, media, chemical, biological, utilities, and political installations and facilities?
- What are the national and local political parties or other groups known to have aims, beliefs, or ideologies contrary or in opposition to those of the U.S.? What are the student, police, military veterans, and similar groups known to be hostile to the U.S.?

6003. CI PLANNING AND THE INTELLIGENCE CYCLE

General

The intelligence cycle is a procedural framework for the development of mission-focused intelligence support. It is not an end in itself, nor should it be viewed as a rigid set of procedures that must be carried out in an identical manner on all occasions. The commander and the intelligence officer must consider each IR individually and apply the intelligence cycle in a manner that develops the required intelligence in the most effective way.

The application of the intelligence cycle will vary with the phase of the planning cycle. In theory, a unique iteration of the intelligence cycle is carried out for each individual requirement. In practice, particularly during the planning phase, requirements are grouped together and satisfied through a single, concurrent intelligence development process that addresses CI requirements. During the planning phase, intelligence development is generally carried out through two major iterations of the intelligence cycle. The first primarily supports decision planning. Completion of this iteration of the intelligence cycle results in the preparation and use of basic intelligence and CI products-intelligence and CI estimates, supporting studies, and IPB analysis-that describe the battlespace and threat. These products form the basis for development and selection of MAGTF COAs. The second iteration of the intelligence cycle supports execution planning. It is an outgrowth of the selection of the COA and formulation of the concept of operations; the implementation of the intelligence collection, production and dissemination plan; refinement of IPB analysis, and the generation of mission-specific intelligence products and CI measures integrated with the concept of operations to support mission execution. During execution, requirements are satisfied on a more individualized basis. New requirements are usually generated in response to a specific operational need. Each requirement is unique and must be satisfied in a timely manner to facilitate rapid decisionmaking and the generation or maintenance of tempo (see figure 6-4).

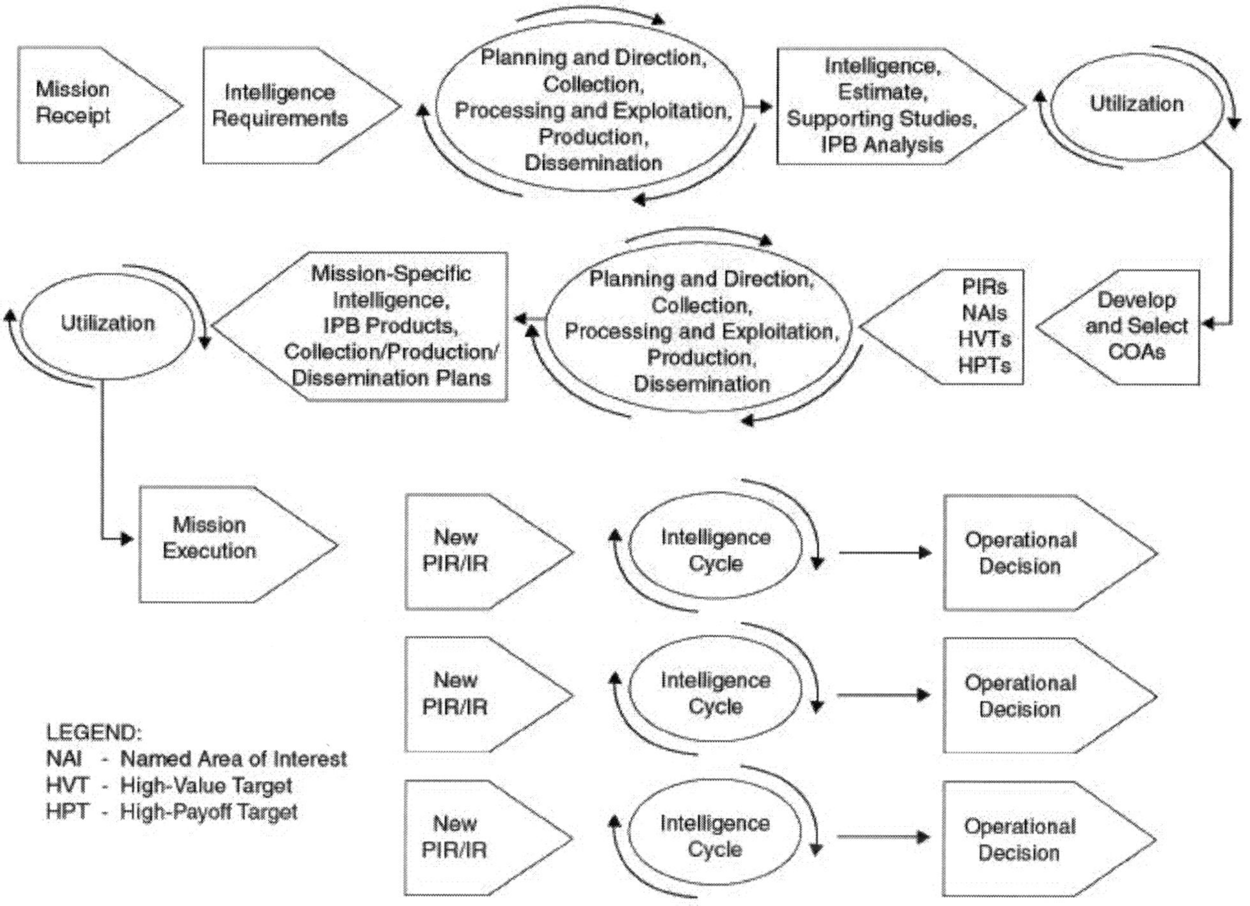

Figure 6-4. Application of the Intelligence Cycle.

The G-2/S-2 provides CI participation and assistance early in the planning phase of tactical operations. Commanders benefit from CI information given at this phase because it helps to formulate tactical plans and because CI/HUMINT operations generally require more time than other intelligence disciplines to yield substantive results. CI also provides the commander with capabilities that are offered by no other discipline or technical system. CI is one of the tools that can help the commander anticipate the action of the enemy. Particular attention is directed to identifying friendly vulnerabilities to be exploited by hostile collection assets and to recommending specific CI measures.

The CI effort focuses on the overall hostile intelligence collection, sabotage, terrorist, and subversive threat. The CI effort is also sufficiently flexible to adapt to the geographical environment, attitudes of the indigenous population, mission of the supported command, and changing emphasis by hostile intelligence, sabotage, terrorist, and subversive organizations.

Planning the Activity

The effectiveness of MAGTF CI operations depends largely on the planning preceding the operation. The G-2/S-2 performs four separate functions in carrying out CI planning responsibilities. First, the effort is directed to collect information on the enemy's intelligence, sabotage, terrorism, and subversion capabilities, coordinating with the ISC, CMDO, and CI/HUMINT Co commander, ensuring CI collection requirements levied upon CI/HUMINT elements are realistic and within the capability of the CI elements and are integrated into the MAGTF's all-source intelligence collection plan. Second, with the ISC, the P&A cell OIC, and the G-3/S-3 force protection planners, the G-2/S-2 supports the production of intelligence on the enemy's intelligence, sabotage, terrorism, and subversion capabilities, including clandestine and covert capabilities. Third, with the ISC and the CMDO, timely dissemination of CI products are assured to MAGTF units. Finally, the G-2/S-2 plans, recommends, and monitors CI measures throughout the command.

CI Collection and Processing of Information

CI elements can collect both CI and intelligence information through the use of overt tactical source HUMINT operations. The determination of CI collection requirements follows the same process and procedures prescribed for other types of IRs (see MCWP 2-1, chapter 3). Especially pertinent to CI planning is information on the enemy's intelligence and reconnaissance capabilities and operations. Included are such matters as the hostile intelligence organization, means available to the enemy for the collection of information, and hostile sabotage, terrorism, and subversive agencies and capabilities.

The area commander normally provides the procedures and authority governing the conduct of overt tactical source HUMINT and certain offensive CI operations. These procedures are covered in detail in the classified addendum, DIAM 58-11, DIAM, and theater collection plans.

CI Collection Sources. CI sources of information include—

- **Casual Sources.** A casual source, by social or professional position, has access to information of CI interest, usually on a continuing basis. Casual sources usually can be relied on to provide information routinely available to them. They are under no obligation to provide information. Casual sources include private citizens, such as retired officials or other prominent residents of an area. Members of private organizations also may furnish information of value.
- **Official Sources.** Official sources are liaison contacts. CI personnel conduct liaison with foreign and domestic CI intelligence, security, and law enforcement agencies to exchange information and obtain assistance. CI personnel are interested in investigative, operational, and threat information.
- **Recruited Sources.** Recruited sources include those who support CFSO and are by design, human source networks dispersed throughout the AO who can provide timely and pertinent force protection information.
- **Refugees, Civilian Detainees, and EPWs.** Refugees, civilian detainees, and EPWs are other sources of CI information. The key to identifying the source of valuable CI force protection information is in analyzing the information being sought and predicting who, by virtue of their regular duties, would have regular, frequent, and unquestioned access to such information.
- **Open Sources.** Open source publications of all sorts (newspapers, magazines, etc.) and radio and television broadcasts are valuable sources of information of CI interest and information. When information is presented in a foreign language, linguist support is required for timely

translation. Depending on the resources, this support can be provided by MAGTF IT personnel, allied personnel, or indigenous employees.

- **Documents.** Documents not openly available, such as adversary plans and reports, are exploited in much the same way as open source publications.

CI Targets. CI targets include personalities, organizations, and installations (PO&I) of intelligence or CI interest, which must be seized, exploited, neutralized or protected. The PO&I targeting triad forms the basis of CI activities. Incidents are also included within CI data bases to conduct trend analysis of potential targets. DIA has the responsibility, in response to a validated requirement, to establish and maintain CI data bases. Operational control of the data base will pass to the JTF HQ, once deployed.

The P&A cell's CI analytical team has the responsibility of establishing and maintaining CI data bases for the MEF and will coordinate with the local collection elements to eliminate duplication of effort and maximize information sharing (within smaller MAGTFs, the CI/HUMINT Co detachment or HST performs this function).

Selecting and assigning targets is based on an assessment of the overall hostile threat, unit mission, commander's intent, designated PIRs and other IRs, and the overarching intelligence and force protection concepts of operations. The assessment considers both the immediate and estimated future threats to security. It normally is conducted at the MAGTF level where the resultant CI target lists are also produced and include any CI targets assigned by higher headquarters. Numerical priority designations are assigned to each target to emphasize the relative importance and value of the target. Designations also indicate the degree of security threat and urgency in neutralizing or exploiting the target. Priority designations established by higher headquarters are not altered; however, lower echelons may assign priorities to locally developed targets.

CI targets are usually assigned priority designations according to the following criteria:

- **Priority One.** Priority One targets represent the greatest security threat to the MAGTF. They also possess the largest potential source of intelligence or CI information/material that must be exploited or neutralized as soon as possible.
- **Priority Two.** Priority Two targets are of lesser significance than priority one targets. They are to be taken under control after priority one targets have been neutralized or exploited.
- **Priority Three.** Priority Three targets are of lesser significance than priority one or two targets. They are to be neutralized or exploited as time and personnel permit.

Personalities. Except for well-known personalities, most persons of CI interest are identified and developed by CI units once operations have commenced. Personalities are divided into three, color-coded categories: threat to security, intentions unknown, and assist the intelligence and CI effort.

- **DETAIN (Black) List.** A CI listing of actual or potential enemy collaborators, sympathizers, intelligence suspects, and other persons whose presence menaces the security of friendly forces (JP 1-02). The black list includes the following persons:

- Known or suspected enemy or hostile espionage, sabotage, terrorist, political, and subversive individuals.
- Known or suspected leaders and members of hostile paramilitary, partisan, or guerrilla groups.
- Political leaders known or suspected to be hostile to the military and political objectives of the United States and/or an allied nation.
- Known or suspected officials of enemy governments whose presence in the theater of operations poses a security threat to the U.S. Forces.
- Known or suspected enemy collaborators and sympathizers whose presence in the theater of operations poses a security threat to the U.S. Forces.
- Known enemy military or civilian personnel who have engaged in intelligence, CI, security, police, or political indoctrination activities among troops or civilians.
- Other enemy personalities such as local political personalities, police chiefs, and heads of significant municipal and/or national departments or agencies.

Gray lists, compiled or developed at all echelons of command, contain the identities and locations of those personalities whose inclinations and attitudes toward the political and military objectives of the U.S. are unknown.

- **OF INTEREST (Gray) List.** Regardless of their leanings, personalities may be on gray lists when known to possess information or particular skills required by friendly forces. They may be individuals whose political motivations require further exploration before they can be used effectively. Examples of individuals who may be included in this category are:
  - Potential or actual defectors from the hostile cause whose credibility has not been established.
  - Individuals who have resisted, or are believed to have resisted the enemy government and who may be willing to cooperate with friendly forces, but whose credibility has not been established.
  - Nuclear, biological, chemical and other scientists and technicians suspected of having been engaged in enemy weapons of mass destruction and other programs against their will.

- **PROTECT (White) Lists.** White lists, compiled or developed at all echelons of command, contain the identities and locations of individuals in enemy controlled areas. These individuals are of intelligence or CI interest. They are expected to be able to provide information or assistance in the accumulation of intelligence data or in the exploitation of existing or new intelligence areas of interest. They are usually in accord with or favorably inclined toward U.S. policies. Their contributions are based on a voluntary and cooperative attitude. Decisions to place individuals on the white list may be affected by the combat situation, critical need for specialists in scientific fields, and such intelligence needs as indicated from time to time. Examples of individuals included in this category are:
  - Deposed political leaders of a hostile state.
  - Intelligence agents employed by U.S. or allied intelligence agencies.
  - Key civilians in areas of scientific research, including faculty members of universities and staffs of industrial or national research facilities whose credibility have been established.

- Leaders of religious groups and other humanitarian groups.
- Other persons who can significantly aid the political, scientific, and military objectives of the U.S. and whose credibility has been established.

Organizations. These include any organization or group that is an actual or potential threat to the security of JTF or allied forces and must be neutralized. However, an organization or group may present a threat that is not immediately apparent. The enemy frequently camouflages his espionage or subversive activities by establishing front organizations or groups. If these organizations are permitted to continue their activities, they could impede the success of the military operations. Examples of hostile organizations and groups of major concern to the CI unit during tactical operations include—

- Hostile intelligence, sabotage, subversive, and insurgent organizations or groups.
- National and local political groups and parties known or suspected to have aims, beliefs, or ideologies contrary or in opposition to those of the United States.
- Paramilitary organizations, including student, police, military/veterans, and ex-combatant groups, known to be hostile to the U.S.
- Hostile sponsored groups and organizations whose objectives are to create dissension and spread unrest among the civilian population in the AO.

Installations. These include any installation, building, office, or field position that may contain information or material of CI interest or that may pose a threat to MAGTF security. Examples of installation type targets are—

- Installations formerly or currently occupied by enemy espionage, sabotage, subversive, or police organizations, including prisons and detention centers.
- Installations occupied by enemy intelligence, CI, security, or paramilitary organizations, including operational bases, schools, and training sites.
- Enemy communication media and signal communication centers.
- Research centers and chemicals laboratories used in the development of weapons of mass destruction.
- Enemy political administrative headquarters.
- Production facilities, supply areas, and other installations to be taken under control to deny support to hostile guerrilla and partisan elements.
- Public utilities and other installations to be taken under early control to prevent sabotage. These installations are usually necessary for the rehabilitation of civil areas under U.S. control.
- Embassies and consulates of hostile governments.

Incidents. The recording of details of incidents occurring within the AO allows for trend analysis that may reveal patterns and indications of future intentions. Incidents do not occur in a vacuum. They are planned, organized, and carried out by individuals acting alone or in groups. Detailed recording of incidents that occur within an AO is a critical ingredient in the analysis of trends and patterns designed to identify indications of future intentions.

Matrix manipulation, link analysis, and visual investigative analysis are tools often used in the incident analysis process. The matrix allows for a considerable amount of information to be stored in a relatively small space. Link analysis then analyzes these bits of information by displaying the links that exist between them. Visual investigative analysis is used to time sequence names and events to show a clearer picture of these relationships.

CI Target Reduction. The timely seizure and exploitation of CI targets requires a detailed and well-coordinated CI reduction plan. This plan should be prepared well in advance and kept current. CI elements supporting tactical assault units normally prepare the reduction plan. Assigned or developed targets located within the unit's AO, are listed in the reduction plan. This plan is based on the MAGTF's scheme of maneuver, with CI targets listed in the sequence in which they are expected to appear in the AO. However, the target priority designations remain as assigned on the CI target list with highest priority targets covered first when more than one target is located in the same general area. Neutralized and exploited targets are deleted from the CI reduction plan and appropriate reports are submitted. A well-prepared and comprehensive CI reduction plan ensures coverage of significant CI targets. It also allows CI units to conduct daily operations based on established priorities. (See appendix D for a sample CI reduction plan.)

CI Measures Worksheet. The CI worksheet is prepared or revised based on the conclusions reached in the intelligence estimate of the enemy capabilities for intelligence, subversion, terrorist activities, and sabotage. This worksheet is an essential aid in CI planning. It is also the basis for preparing CI orders and requests. (See appendix D for a sample CI measures worksheet.)

CI Analysis and Production

Analysis and production is the heart of CI despite the quality and quantity of information is gathered, it will be worthless if is not turned into intelligence and disseminated to commanders and planners in time to use for decisionmaking. Critical to the success of MAGTF CI activities is taking collected information and producing tactically relevant intelligence (usually via all-source intelligence production), and providing it to commanders and planners in a timely manner. The transition of raw information into finished intelligence is the process of analysis. Fusing the finished intelligence into something usable by the customer is known as production.

Offensive

- Targeting for fire and maneuver.
- Physical security.
- Counterreconnaissance.
- Countersabotage.
- Penetration and exploitation operations.

Defensive

- Deception operations (OPSEC).
- Counterespionage operations.
- Information security.
- Personnel security. Counterterrorism

CI Production Threat Focus. CI analysis and production is focused on three well-defined threat activities: HUMINT, IMINT, and SIGINT.

- **Counter Human Intelligence (C-HUMINT).** C-HUMINT requires effective and aggressive offensive and defensive measures. Our enemies collect against MAGTFs using both sophisticated and unsophisticated methods. We must combat all of these methods to protect our force and to ensure the success of our operations. MAGTF CI elements recommend countermeasures developed by CI analysts that the commander can take against enemy collection activities. CI C-HUMINT analysis focuses not only upon the standard enemy CI targets within the AO, but also upon the intelligence product most likely being developed through their collection activities. The analytical effort should attempt to identify the enemy's HUMINT cycle (collection, analysis, production, targeting) and key personalities. To produce a complete and accurate analysis and then develop effective products, CI analysts may need access to considerable data and require significant resources. CI analysts will require collection in the areas of subversion, espionage, sabotage, terrorism, and other HUMINT supported activities. Collection of friendly force data is also required to substantiate CI analytical findings and recommendations. Consistent with time, mission, and availability of resources, efforts must be made to provide an analytical product identifying the enemy's intelligence, espionage, subversion, sabotage, and terrorism efforts. To accomplish C-HUMINT, MAGTF CI elements will conduct CI investigations, collections, and operations. Key C-HUMINT tasks include—

- Identifying the hostile HUMINT collectors and producers.
- Developing, maintaining, and disseminating multi-discipline threat data and intelligence on organizations, locations, and individuals of CI interest. This includes insurgent and terrorist infrastructure and individuals assisting in the CI mission. It also includes performing personnel security investigations and records checks on persons in sensitive positions and whose loyalty is questionable. Also, personnel must be educated and trained in all aspects of command security. A component of this is the multi-discipline threat briefing. Briefings can and should be tailored, both in scope and classification level. Briefings can then be used to familiarize MAGTF units with the nature of the multi-discipline threat posed against the command or activity. Additionally, CI elements must search for enemy personnel who pose an intelligence collection or terrorist threat to the MAGTF. Should CI investigations result in identifying the location of terrorists, their apprehension is done with military, civil, and law enforcement authorities. Also, debriefing of selected personnel (friendly and hostile) including combat patrols, aircraft pilots, or other elements possessing information of CI interest is necessary.
- Neutralizing or exploiting these to deny the enemy key friendly force information.
- Controlling our own information and indicators of operations so they are not readily accessible to the enemy's HUMINT operations.
- Supporting C-HUMINT commanders through effective and stringent adherence to physical, information, and personnel security procedures. They apply force or assets to ensure security daily. MAGTF CI and intelligence elements provide continuous and current threat information so the command can carry out its security responsibilities.

See appendix C, section I, for additional information on C-HUMINT.

- **Counter Imagery Intelligence (C-IMINT).** C-IMINT requires CI analysts to have an in-depth knowledge of the supported commander's plans, intentions, and proposed AO as far in advance as possible. CI analysts must have access to available data and intelligence on enemy IMINT methodology, systems, and processing as well as in-depth information on commercial satellite systems and the availability of their products to the enemy or to other supporting parties. CI analysts attempt to define the specific imagery platform deployed against the MAGTF and the cycle involved (time-based) from time of imaging through analysis to targeting. Knowledge of enemy intelligence support to targeting is critical in developing countermeasures to defeat, destroy, or deceive enemy IMINT. For ground-based HUMINT oriented IMINT (video cassette recorders, cameras, host nation curiosity, news media organizations), MAGTF CI elements will be required to collect the data for analysts. This type of information cannot be reasonably considered to exist in any current data base. Traditional enemy IMINT data is readily available and should not require any CI collection effort. However, collection to support CI (overflights of friendly forces by friendly forces) during identified, critical, and IMINT vulnerable times will validate CI C-IMINT findings and support countermeasures planning and execution. This will be of immense value to the CI analyst and the supported commander in determining what enemy imagery has been able to exploit. (See appendix C, section II, for additional information on C-IMINT.) The enemy may possess or acquire IMINT systems or products with comprehensive and sophisticated capabilities. The MAGTF must have in place carefully developed countermeasures to negate any tactical and strategic threat. The enemy may acquire IMINT through a variety of ways, from handheld cameras to sophisticated satellite reconnaissance systems. Such IMINT capabilities may include—

- Aerial cameras.
- Infrared sensors.
- Imaging radars.
- Electro-optical sensors (TV).
- Multispectral and digital imagery products.

Offensive

- Targeting for fire and maneuver.
- Electronic attack.

Defensive

- OPSEC countermeasures.
- Use of secure telephone.
- Signals security procedures.
- Deception operations.

- **Counter Signals Intelligence (C-SIGINT).** C-SIGINT operations, including COMSEC monitoring and information systems security, are conducted during peace, war, and MOOTW to enhance MAGTF force protection, survivability, mobility and training; provide data to identify friendly CIS vulnerabilities; develop countermeasures recommendations and plans; and when implemented, determine if countermeasures are effective. C-SIGINT includes full identification of the threat and an integrated set of offensive and defensive actions designed to counter the threat. Counter-SIGINT focuses upon the enemy's entities that can conduct SIGINT and EW against friendly forces. It also focuses on the intelligence that is most likely being collected and produced from their efforts. C-SIGINT analysis effort should be fully automated (data storage, sorting, and filing). CI analysts require SIGINT data collection to support vulnerability assessment and countermeasure evaluation. Validation of vulnerabilities (data and operations that are exploitable by the enemy's SIGINT operations) and the effectiveness of implemented countermeasures (a before and after comparison of MAGTF electromagnetic signatures and data) will be nearly impossible without active and timely collection as a prerequisite to analysis. CI analysts require a comprehensive data base consisting of enemy SIGINT systems, installations, methodology, and associated SIGINT cycle information. Friendly CIS systems and user unit identification must be readily available, as well as a library of friendly countermeasures and a history of those previously implemented countermeasures and results achieved. CI analysts should, at any given time, be able to forecast enemy SIGINT activity. However, such estimates must rely upon other CI, SIGINT, and IMINT collection and access to adjacent friendly unit CI files. Information on enemy SIGINT must be readily accessible from intelligence elements higher as well as lower in echelon than the supported command. Effective conduct of C-SIGINT requires close coordination and integrated production between MAGTF CI, SIGINT and all-source intelligence producers. C-SIGINT provides commanders and planners with the knowledge to assess the risk and probable success of alternative courses of action before a plan is implemented. C-SIGINT is a cyclic process requiring a strong analytical approach integrating MAGTF CI, SIGINT, CIS and force protection personnel. C-SIGINT is based on a thorough knowledge of—

- Enemy SIGINT capabilities and tactics, techniques and procedures.
- MAGTF and other friendly forces' communications and information systems profile.
- Enemy operations and plans.
- Realistic security measures, both INFOSEC and physical, that can be taken to deny information to the enemy. (See appendix C, section III, for additional information on C-SIGINT.)

CI Analytical and Production Functions. CI analysts perform the following analytical and production functions:

- Analyze the multi-discipline intelligence, espionage, subversion, sabotage, and terrorism threats targeted against the MAGTF.
- Assess enemy intelligence vulnerabilities and susceptibilities to friendly deception efforts and other countermeasures.
- Support MAGTF force protection vulnerability assessment.
- Develop, evaluate, and recommend countermeasures to reduce, eliminate, or take advantage of MAGTF vulnerabilities.
- Support rear area operations by identifying intelligence, espionage, subversion, sabotage and terrorism threats to rear area units and installations (to include low-level agents responsible for sabotage and subversion).
- Nominate CI targets for exploitation, neutralization, or destruction.
- Develop and maintain a comprehensive and current CI data base.
- Identify CI IRs and provide these to the collection officer.

CI Products. CI products convey pertinent intelligence resulting from CI analysis to the commanders and planners in a readily useable form. CI analysts prepare a range of products; some focused upon specific needs and others of a more general nature. Among these products of most use to commanders and planners are CI estimates, surveys/vulnerability assessments, summaries, and threat assessments.

- **CI Estimate.** Within the MAGTF a CI estimate is normally prepared only by the ISC for the MAGTF G-2/S-2 and disseminated throughout the MAGTF. The CI estimate forms the basis of the CI plan and operations. It includes the enemy's capabilities and limitations for intelligence, subversion, terrorism, sabotage, and the effects of the characteristics of the area on these capabilities and friendly CI measures. If a CI estimate is not prepared, such CI planning information can be consolidated within the basic intelligence estimate (appendix 11 to annex B, Intelligence). Key parts of the CI estimate include sections on enemy intelligence, subversion, sabotage, guerrilla warfare, terrorism, and the effects of the area on these enemy capabilities. See appendix C for a sample format of a MAGTF CI estimate.
- **CI Survey/Vulnerability Assessment.** CI surveys/vulnerability assessments are studies conducted to provide a supported command or

agency a picture of its susceptibility to foreign intelligence collection. The CI survey/vulnerability assessment assesses a unit's security posture against threats detailed in the CI estimate. The survey should identify vulnerabilities to specific hostile intelligence, espionage, sabotage, subversion or terrorist capabilities and provide viable recommendations to eliminate or minimize these vulnerabilities. The survey should be as detailed as possible. It will usually be prepared by the CI/HUMINT Co, drawing on the expertise of other disciplines such as engineers, military police, civil affairs, and other intelligence and reconnaissance personnel. The survey must look forward, in both space and time, to support the development of CI measures necessary to protect the unit as it carries out successive phases of the operation. The objective of a CI survey/ vulnerability assessment provides commanders a realistic tool to evaluate internal force protection or security programs, and to provide a decisionmaking aid for the enhancement of these programs. CI surveys/ vulnerability assessments include—

- Estimating the enemy's likely PIRs and then evaluating the enemy's and any potential supporting forces' multi-discipline intelligence collection and production capabilities to answer these. Identifying MAGTF and other friendly forces' activity patterns (physical and electronic), friendly physical and electronic signatures, and resulting profiles to enhance OPSEC.
- Monitoring or collecting MAGTF and other friendly forces' CIS transmissions to aid in vulnerability assessments, and providing a more realistic and stable basis to recommend countermeasures. (Note: these operations are generally conducted either by elements of the radio battalion or other supporting elements.)
- Identifying MAGTF and other friendly forces' vulnerabilities based upon analysis of collected information and recommendations of countermeasures.
- Analyzing the effectiveness of implemented countermeasures. See appendix D for a CI survey checklist and the format for a CI survey/vulnerability assessment.

- **Counterintelligence Summary (CISUM).** CISUM is a graphic portrayal of the current situation from a CI point of view. It will be prepared either by the P&A cell or the CI/HUMINT Co's operations/analysis element. CI analysts use the CISUM to show known adversary intelligence units as well as known/estimated threats within the friendly area. CISUM is a periodic report usually covering a 12-hour period. It shows friendly targets identified as adversary objectives during the specified timeframe. CI analysts include a clear, concise legend on each CISUM showing the time period, map reference, and symbols identifying friendly and adversary information. As analysts identify a friendly critical node, element, or resource as an adversary combat or intelligence collection target, they put a box around it and label it with a T number. The legend explains whether the T is—
  - An enemy intelligence target.
  - A source and time confirmation.
  - An enemy resource or element that will attack or collect against the target in the future.
  - The expected timeframe for the enemy to exploit the target.
  - The CISUM might portray the following information:
  - Satellite or tactical reconnaissance patterns over the AO.

- Sweeps by enemy side looking airborne radar (SLAR) or EA air platforms to the full extent of their maximum ranges.
- Suspected landing zones or drop zones that will be used by an enemy element in the rear area.
- Area or unit that has received unusual enemy jamming or other electronic attacks. Movement of an enemy mobile SIGINT site forward along with a graphic description of the direction and depth of its targeting.
- Location of an operational enemy agent or sabotage net.
- Last known location of threat special operations forces.

- **CI Threat Assessment.** The CI threat assessment (see page 6-18) is a four-paragraph statement which is published as often as necessary or when significant changes occur, depending on the situation and the needs of the commander. The CI threat assessment provides justification for CI target nominations and guidance for CI production. It will generally be produced through a combined effort of the P&A cell and the CI/HUMINT Co's operations/analysis element. Essentially, the CI threat assessment provides the following:
  - A quick overview of significant activity during the reporting period.
  - An assessment of the intelligence damage achieved by the enemy.
  - A projected assessment of enemy activity for the next reporting period.
  - CI target nominations.

CI Dissemination

Refer to chapter 5 for a detailed review of CI dissemination planning considerations.

6004. CI PLANNING REQUIREMENTS AND CONSIDERATIONS

The following describes the CI planning requirements, considerations, and activities. It is provided as a guide for MAGTF intelligence, force protection, and CI personnel in planning MAGTF operations.

Formulation of the Commander's Estimate

CI planning activities follow a logical sequence consistent with the MCPP and the six functions of intelligence.

- Provide assistance to the command operations security program. During the initial planning phase, CI assets provide assistance to the G-3/S-3 in establishing force protection planning and operations.
- Complete studies of the enemy organization, weapons and equipment, techniques, and effectiveness in conducting intelligence, subversion, terrorism, and sabotage operations. Timely completion and dissemination throughout the MAGTF of the CI estimate is critical.
- Complete the CI survey/vulnerability assessment to assist with MAGTF force protection planning and countermeasures development and implementation.
- Release CI planning information and products per PIRs and IRs and specified reporting criteria established by the CMDO or as directed by the G-2/S-2. It is critical that CI personnel follow-up with recipients of these products to ensure information is understood and to identify early any resulting new CI IRs.

COUNTERINTELLIGENCE THREAT ASSESSMENT

1. (U) Enemy Activity During Period___ to ___

a. (U) HUMINT: Summarize in one paragraph all know HUMINT activity during the reporting period. Compile data from HUMINT situation overlay, matrices, link diagrams, and other CI products.

b. (U) SIGINT: Summarize in one paragraph all known SIGINT activity. Compile from SIGINT situation overlay, matrices, and other CI products.

c. (U) IMINT: Summarize in one paragraph all known IMINT activity. Compile from IMINT situation overall, pattern and analysis chart, and other CI products.

d. (U) OTHER: Summarize any other pertinent enemy activity not already addressed.

2. (U) Counterintelligence Damage Assessment for the Period ___ to ___ (list DTGs)

a. (U) Briefly assess the CI damage to MAGTF units for whom the assessment is being prepared. Assessment is based upon enemy intelligence and reconnaissance activities that were identified, analyzed, and reported, measured against the MAGTF operations profile and countermeasures implemented. Coordination with G-3/S-3 OPSEC/force protection staff is essential when preparing this paragraph.

3. (U) Estimated Enemy Activity for the Period ___ to ___ (list DTGs)

a. (U) HUMINT: briefly describe estimated enemy HUMINT activity for the reporting period.

b. (U) SIGINT: briefly describe estimated enemy SIGINT activity for the reporting period.

c. (U) IMINT: briefly describe estimated enemy IMINT activity for the reporting period.

d. (U) OTHER: briefly describe any other pertinent estimated enemy activity not addressed above.

4. (U) Counterintelligence Target Nominations

a. (U) EXPLOITATION: identify any CI targets worthy of exploitation. Provide recommended timeframe, methods of exploitation, location, justification, and any other pertinent information.

b. (U) NEUTRALIZATION: identify any CI targets worthy of nertralization. Provide recommended timeframes, methods of neutralization, locations, justifications, and any other pertinent information.

c. (U) DESTRUCTION: identify any CI targets worthy of destruction. Provide recommended timeframe, methods of engagement, locations, justifications, and any other pertinent information.

- Identify necessary restrictions on informing MAGTF personnel about mission details, D-day, H-hour, designated landing beaches, helicopter landing zones, selected objective and other critical friendly force information requirements.
- Coordinate with the G-6/S-6 and subordinate commanders, and provide assistance with MAGTF communications and information systems security.

Support to Targeting

Supervise the accomplishment of CI operations in accordance with the CI plan. Including—

- Exploit sources of information to provide critical intelligence to commanders and planners. Provide pre-targeting surveillance and route reconnaissance enabling the commander to determine the appropriate method and force to be applied against the target.
- Assist with identifying MAGTF security and target vulnerabilities, evaluating the relative importance of each (MAGTF targets may be personalities, organizations, installations or capabilities). Additionally, identify security vulnerabilities of nongovernmental organization, private volunteer organizations, media, and other such organizations that are within the MAGTF's AO.
- Locate and recover contraband materials, such as arms, explosives, communication equipment, food, medical supplies, or other items not surrendered in accordance with proclamations. This denies critical capabilities to adversaries.
- Seize, exploit, and protect CI targets.

Combat Assessment

Continue an aggressive CI/HUMINT collection program in response to the MAGTF PIRs and IRs and in protection of MAGTF EEFIs to gauge the impact of friendly actions on the enemy and civilian populace and to evaluate the effectiveness of MAGTF security countermeasures.

6005. CI PLANS AND ORDERS

General

See appendix F for a list of MAGTF CI planning actions associated with each step of the MCPP.

Guidance for the conduct of MAGTF CI operations comes from many sources. The DIA 58 series of manuals and JP 2-01.2, are the principal references for U.S. CI operations and contains policy, direction, guidance, and instruction on how to perform the CI operations, activities and functions in compliance with national directives and security requirements (see appendix H for a detailed listing of CI and related references). Additionally, since MAGTFs will normally be part of a JTF or naval expeditionary force, reference to pertinent combatant command, JTF and fleet orders, guidance, and CI TTPs are necessary to identify unique operating concepts and methodologies and support procedures and formats. MAGTF CI plans and orders are prepared by the G-2/S-2. Plans developed by the G-2 plans officer

The G-2 plans officer coordinates the overall initial CI planning effort with the assistance of the ISC, CIHO, and the CO/OICs of organic and supporting CI units.

will then transition to the ISC, who is then responsible for detailed development of CI plans, their integration with other MAGTF intelligence, operations, CIS and logistics plans, and then principal oversight of execution. MAGTF CI plans and orders appear as an appendix to the intelligence annex of the MAGTF operation plan or order and will focus on internal MAGTF CI requirements, operations and TTP.

The CI Appendix

The CI appendix to an operations plan OPLAN or OPORD will be prepared consistent with format outlined in the Joint Operational Planning and Execution System (JOPES) and appear as appendix 3 (CI Operations) to Annex B (Intelligence) in all operations plans and orders. (See appendix B for a sample CI appendix format.) The CI appendix should include—

- Friendly forces to be used including—
  - Personnel augmentation requirements.
  - CI units of adjacent or other theater forces and the support expected.
  - Joint force maritime component commander (JFMCC) and ATF CI elements that may provide support to the landing force in amphibious operations.
  - Pertinent CI capabilities and support from the combatant command's joint intelligence center/joint analysis center, JTF joint Force J-2 (HUMINT staff element (J-2X), joint force land component commander, joint force air component commander, and other component commanders/task forces within JTF operations.
- Planned arrangement, employment, and use of external CI support including any special collection, production, dissemination, and CIS arrangements.
- Coordinating instructions established for the planning and control of CI operations including technical support expected from higher headquarters.
- MAGTF CI elements tasking.
- CI production priorities and plans.
- CI dissemination priorities and plans, including communication and information systems support to the MAGTF CI effort.
- CI unique equipment and logistics requirements.
- An appendix providing MAGTF countersigns, challenges, passwords, and supporting procedures.

CHAPTER 7. EXECUTION OF CI ACTIVITIES

7001. MAGTF CI OPERATIONS

The MAGTF CI mission and concept of operations depend on many factors. Major external factors affecting CI operations include the mission, commanders' intent, and the size and nature of the AO and AOI, operations and intelligence concept of operation, stated PIRs, EEFIs, C2 of the MAGTF, and the JTF. Key internal factors include the type of C2, various potential employment options, communication and information systems capabilities, and unique CI equipment. Finally, key tasks, such as completing necessary CI surveys/vulnerability assessments, the CI estimate, and development of a CI target-reduction plan, all affect MAGTF CI employment and operations.

Planning

The successful accomplishment of the command's mission requires thorough planning by the commander. The following must be considered when planning for CI activities:

- Intelligence concept of operations, designated PIRs, IRs, EEFIs, and supporting collection, production, and dissemination plans.
- Force protection concept of operations and designated EEFIs.
- Detailed study of available maps and photographs of potential areas of operations.
- Study of available intelligence products about the AO and threat.
- Location and plotting of critical CI targets, categorized by personalities, organizations, and installations. These are categorized and studied and plans formulated for coverage and reduction. Target priorities must be assigned in advance to ensure efficient use of personnel. The area surrounding a target is studied to determine points where sealing off would be most effective. Streets and approaches to targets are studied thoroughly, thereby minimizing the need for extensive physical reconnaissance. Main traffic routes are studied to determine locations in which to establish screening centers and checkpoints.
- Acquisition or development of personalities (black, gray, and white lists), organizations, installations, and incident data bases (POI&I) for the target area.
- Study of all available records to identify host country, third party or other officials and leaders within the AO that the enemy is hostile to. Also study of other persons who could be of value in administrative assignments. These include members of the local police force, fire department, post office, railway, telephone, telegraph, and broadcasting stations. Much of this data can be obtained from the U.S. country team, civil affairs units, and other sources.
- Acquisition of available information on pro-American or anti-opposition elements. These elements include guerrillas and partisans in the zone of operations and other areas that would facilitate immediate use of such groups if necessary.

- Acquisition and study of all information on other underground forces, groups, and personnel who, by reason of training and experience, can provide assistance in the conduct of CI interrogations.
- Other sources of information and intelligence such as CI contingency materials (CICM), MDITDS, DCIIS, and the various all-source intelligence data bases. CICM are focused CI analytical products such as sanitized mapping, imagery, and reference material available from the P&A cell's CI analytical team, the combatant command JIC's CI analytical cell through the combatant command's CI Staff Officer, and DIA's Operational Intelligence Coordination Center, CI Division and Transnational Threat Division.

Command and Control

See paragraphs 3002, 3003, and 5002 for detailed discussion of C2.

MAGTF CI operations will generally be conducted in GS of the MAGTF. Tactical and technical control of MAGTF CI activities is generally centralized under the intel bn and CI/HUMINT Co commanders as directed by the MAGTF G-2/S-2. CI/HUMINT Co direct support or attachment may be necessary in the following situations:

- When subordinate units' missions requires organic CI support.
- When centralized MAGTF control is unfeasible.
- When the operational tempo is high.

Tactical Deployment

During both static and fluid tactical situations in populated areas, CI/HUMINT Co CP normally is centrally located and easily accessible to indigenous personnel.

The CP is located to provide maximum assistance to other agencies and to ensure protection by them if required. During high tempo operations, however, the CI/HUMINT Co CP will be located near the IOC (at the MEF CE level) or, in the case of company detachments, the supported unit's main command echelon. When possible, it will be located outside of key vital area perimeters to enhance security while remaining accessible to key indigenous personnel.

In deploying CI personnel, consideration is given to retaining at least one CI team at the headquarters for special assignments and emergencies.

When CI elements are held in reserve, personnel are organized and equipped so that the augmentation subteams may be immediately dispatched to forward units that require CI support or reinforcements.

CI elements are attached to or placed in direct support of subordinate units sufficiently in advance to coordinate operational, intelligence, and CIS and CI plans supporting the units' mission, IRs, and concepts of operations.

7002. CI SCREENING OPERATIONS

CI screening operations are designed to identify and apprehend enemy intelligence agents, subversives, terrorists, and saboteurs attempting to infiltrate friendly lines or conceal themselves among the population. The purpose of CI screening operations is to identify persons of CI interest and gather information of immediate CI interest.

During conventional combat situations, screening operations primarily consist of screening refugees and EPWs at mobile and static checkpoints in populated areas in cooperation with other MAGTF elements such as military police, ITs, civil affairs (CA), combat units, and psychological operations teams.

Persons of CI Interest

The following are examples of categories of persons of CI interest:

- Persons suspected of attempting to infiltrate through refugee flow.
- Line crossers.
- Deserters from enemy units.
- Persons without identification papers or with forged papers (inconsistent with the norm).
- Repatriated prisoners of war and escapees.
- Members of underground resistance organizations seeking to join friendly forces.
- Collaborators with the enemy.
- Target personalities, such as those on the personalities list (black, gray or white lists).
- Volunteer informants.
- Persons who must be questioned because they are under consideration for employment with U.S. forces or for appointment as civil officials by CA units.

Coordination

CI personnel plan these screening operations with the following:

Commander

The commander is concerned with channeling refugees and EPWs through the AO, particularly in the attack, to prevent any hindrance to unit movement or any adverse effect on unit mission. Accordingly, screening operations must be compatible with the supported commander's concept of operations and scheme of maneuver.

ITs

MAGTF IT personnel must understand what CI is looking for and have the commander's current PIRs, IRs, and EEFIs. Close coordination with interrogators is essential for successful CI operations.

Military Police

Military police (MP) elements are responsible for collecting EPW and civilian internees from capturing units as far forward as possible in the AO. MP units guard the convoys transporting EPW and civilian internees to EPW camps and command and operate the EPW camps.

Civil Affairs

CA elements are responsible for the proper disposition of refugees.

Psychological Operations
Psychological operations (PSYOP) elements, under the G-3's cognizance, contribute to screening operations by informing the populace of the need for their displacement.

Local Civil Authorities in Hostile Areas
Civil authorities in hostile areas are included in planning only if control has been returned to them.

Preparation

Prior to the operation, CI personnel must become thoroughly familiar with all available information concerning the enemy intelligence organization, the military and political situation within the enemy controlled area, and the geography of the area.

Enemy's Intelligence, Infrastructure, and Organization
To successfully identify enemy intelligence agents, CI personnel must be knowledgeable of the enemy intelligence organization, including its mission, methods of operation, officials, schools and training, known agents, equipment, policies, and regulations.

Regulations
Knowledge of the political situation and of the restrictions placed on the population within the enemy controlled area aid in detecting discrepancies during the screening. Information required includes travel restrictions, curfews, draft and conscription regulations, civilian labor forces and work patterns, and the education system.

OOB
However, CI personnel must be aware of the enemy military units operating within the area. They must also be knowledgeable of their disposition, composition, activities, training, equipment, history, and commander's personalities. This information aids in identifying military intelligence personnel or other persons attempting to hide their identity.

Researching, analyzing, and producing OOB information is primarily the responsibility of the ISC and executed by the P&A cell. Collection of OOB information from human sources is the primary responsibility of the ITs.

AO
CI personnel must also be familiar with the geography and the political, social, and economic conditions of the area. Knowledge of travel conditions, distances, major landmarks, customs, and composition of the population is essential to successful screening operations.

Lists and Information Sheets
CI elements should distribute apprehensions lists and information (or basic data) sheets listing indicators of CI interest to the combat units, MPs or other personnel assisting with the screening operation. Basic data sheets should be tailored to the mission. The basic data sheets are filled out by CI personnel to aid in determining the individual's knowledge, to formulate questions for further interrogation, and are provided to the individuals to be screened, requiring them to record personal data. This form aids in formulating the type of questions to be asked and determines the information needed to satisfy PIRs, IRs, and EEFIs. The Geneva Conventions do not require all of

this. If the person refuses to give the information, there is nothing that can be done about it. Prepare the form in the native language of the host nation and enemy force, if different, and ensure that it is prepared in the proper dialect of the language. Include the following data and anything else judged necessary on the form:

- Full name, aliases, date and place of birth, current and permanent residences, sex, race, religion, marital status, and current citizenship.
- Full name, aliases, date and place of birth, current and permanent residences, sex, race, religion, marital status, and current citizenship of the father, mother, and siblings, including the occupation and whereabouts of each.
- Names of spouse (including female maiden name), date, place of birth (DPOB), nationality, occupation, and personal data on spouse's family, if married.
- Individual's education and knowledge of languages.
- Political affiliations and membership in other groups or organizations.
- Details of the individual's career including schools, military service, technical and professional qualifications, political affiliations, and foreign travels.
- Details of current travel to friendly lines/point of capture, including point of departure, destination, times, and purpose.
- Additional questions may be included that relate to specific indicators revealing areas of CI interest.

Initial Screening

The initial screening is designed to identify those persons who are of CI interest and who require interrogation by CI personnel. EPWs and refugees normally enter EPW and refugee channels rearward of the forward line of own troops for further movement to EPW collection points and compounds in rear areas. Unit intelligence personnel, interrogators or CI personnel usually perform the initial screening.

Initial screening is conducted as soon as possible after the EPWs or refugees come under friendly control.

In the case of a large number of refugees, military police, civil affairs units, psychological operations personnel, and tactical troops, if available, may provide assistance with initial screening.

Persons identified or suspected to be of CI interest are separated from other EPWs or refugees. After information of immediate tactical value has been obtained from personnel of CI interest, they are referred to CI personnel for interrogation. Personnel of CI interest are exploited, if possible. Then rear area CI elements evacuate them to higher headquarters for further detailed interrogation and exploitation. Further CI screening also continues for other EPWs and refugees at the higher echelons.

Procedures for the handling of captured enemy personnel and civilian detainees are contained in MCRP 11.8C, *Enemy Prisoners of War and Civilian Internees*.

Conduct of the Screening

The success of a screening operation is influenced by the degree of preparation and the quality of the information provided to CI and other personnel conducting the initial screening. CI interrogation is the method used to confirm or to deny that the person is of CI interest and to exploit the information obtained, when appropriate. CI interrogation is used throughout the entire screening process.

In many cases, numerous EPWs and refugees preclude CI interrogation of every individual. Those persons who are of CI interest are evacuated through CI channels for further interrogation and exploitation by rear area CI elements. During the conduct of the screening process, persons who are determined not to be of CI interest are returned to EPW or refugee channels as appropriate. A screening or an interrogation report is completed on each individual referred for further interrogation. This report clearly identifies areas of CI interest. It includes as much information as possible concerning the individual's identity and documentation, background, recent activities, and route of travel to friendly lines or point of capture.

CI Screening Report

Appendix D contains formats for the CI screening report and for an interrogation report. The CI screening report should include the following:

Identity

Screen all identifying documents in the form of ID cards, ration cards, draft cards, driver's license, auto registration, travel documents, and passport. Record rank, service number, and unit if a person is, or has been a soldier. Check all this information against the form previously filled out by the detainee if this was done.

Background

The use of the form identified earlier will aid in obtaining the information required; however, certain information areas on the forms will have to be clarified, especially if data indicate a suspect category or the person's knowledge of intelligence information. If the form has not been filled out at this point, try to gain the information through questioning.

Recent Activities

Examine the activities of persons during the days before their detainment or capture. What were they doing to make a living? What connection, if any, have they had with the enemy? Why were they in the MAGTF's area? This line of questioning may bring out particular skills such as those associated with a radio operator, linguist or photographer. Make physical checks for certain types of calluses, bruises or stains to corroborate or disprove their story. Sometimes soil on shoes will not match soil from the area they claim to come from.

Discrepancies in travel time and distances can be the key to the discovery of an infiltrator with a shallow cover story.

Journey or Escape Route

CI personnel should determine the route the individual took to get to MAGTF lines or checkpoints. Question the individual further on time, distance, and method of travel to determine whether or not the trip was possible during the time stated and with the mode of transportation used. By determining what an individual observed enroute, the screener can either check the person's story or pick up intelligence information concerning the enemy forces. ITs are well trained in this process and should be called upon for assistance and training.

Indicators

Indicators aid with identifying possible hostile infiltrators or other targets of CI interest. They are determined after a thorough study of the enemy area, the political and military situation, and the enemy intelligence organization.

For maximum effectiveness, indicators must relate to designated PIRs, EEFIs, and other IRs tasked to CI elements. The following general indicators may serve as a guide to identify persons as possible infiltrators:

- Persons of military age who are not members of the armed forces.
- Persons without identification or with unusual or altered documents.
- Persons attempting to avoid detection or questioning, or displaying peculiar activity.
- Persons using enemy methods of operation.
- Persons possessing unusually large amounts of money, precious metals or gems.
- Persons traveling alone or in pairs.
- Persons having a pro-enemy background, family members in enemy area or who have collaborated with the enemy.
- Persons with a suspicious story, who display any peculiar activity or who have violated regulations in enemy areas.
- Persons having technical skill or knowledge.

Other Methods of Screening

In addition to interrogation, the following methods of screening EPWs and refugees can be used separately or in combination.

- Insertion of informants into EPW compounds and camps, civilian internee camps or into refugee centers.
- Use of concealed informants at screening collection points.
- Use of technical equipment (audio and visual) in holding areas.
- Polygraph examination.
- Specialized identification equipment.

Mobile and Static Checkpoints

Checkpoints are employed in screening operations in populated areas and along routes of travel. Checkpoints are used to detect and prevent enemy infiltration of espionage, sabotage, terrorist, and subversive agents. They are also used to collect information that may not otherwise be available to intelligence units.

Checkpoints are established at key locations throughout the AO, where sufficient space is available for conducting searches and assembling the people to be screened. Provision is made for the security of the checkpoint, and personnel are positioned to the front and rear of the checkpoint to apprehend anyone attempting to avoid it. Figure 7-1 on page 7-8 depicts a typical checkpoint.

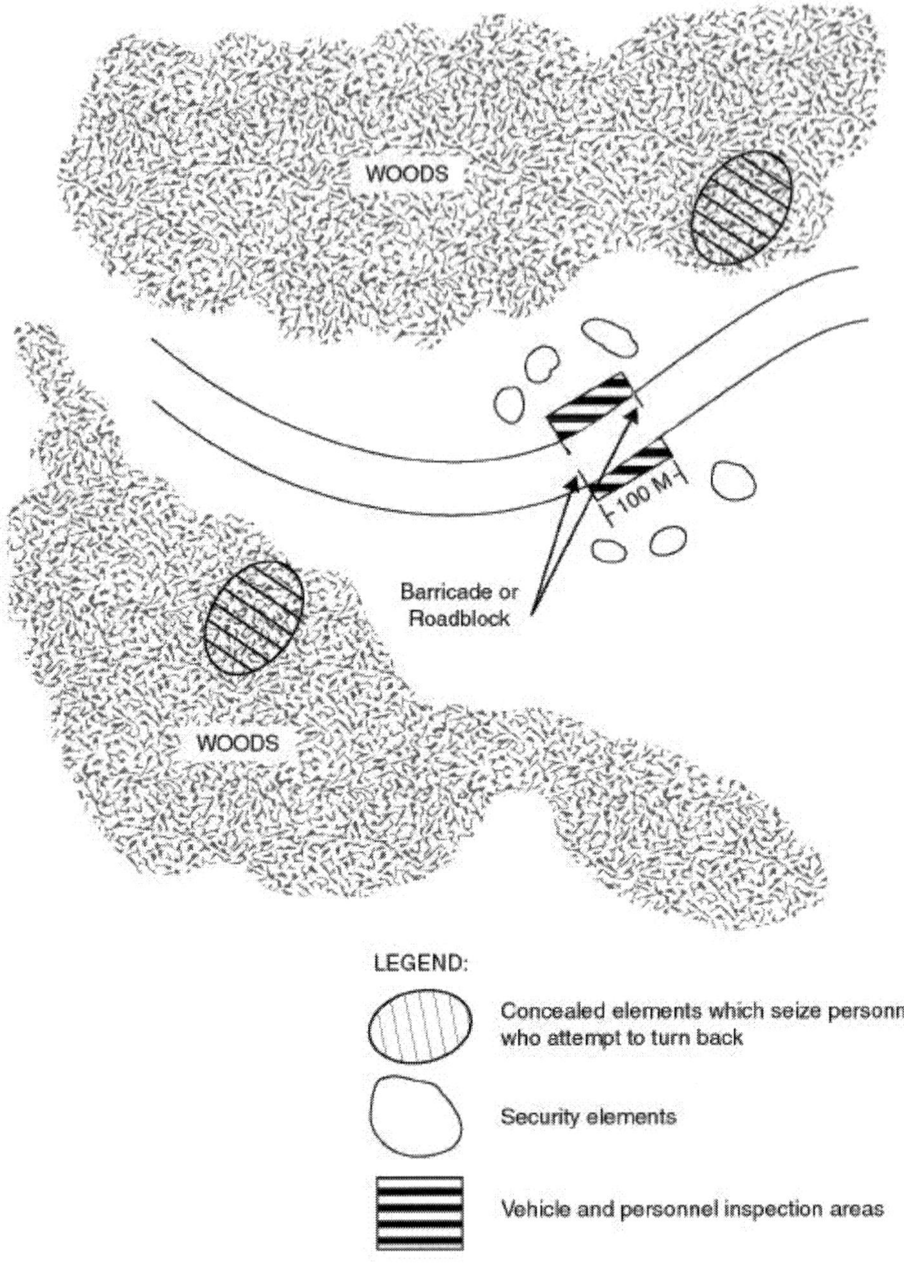

Figure 7-1. Example of a Checkpoint.

There are two types of checkpoints employed in a screening operation—mobile and static.

A mobile checkpoint can be used as a moving system. This system consists of the screening team, either mounted in vehicles or on foot, selecting individuals to be stopped for questioning and a check of identity. The mobile checkpoint also may be established at various locations, usually for periods not to exceed one day.

Static checkpoints are those manned permanently by military police or combat troops at entrances to towns, bridges, and other strategic locations.

The preparation for employment of mobile and static checkpoints is the same as for other screening operations. Lists of persons known or suspected of enemy activity (black—detain and gray—of interest lists) and lists of indicators are normally used in the screening operation. Specialized detection equipment (e.g., metal or explosive detectors) may also be used, if available.

Screening teams may be composed of combat troops, intelligence interrogators, military police, CI personnel, civil affairs personnel or a combination of such personnel. Screening teams conduct the initial screening and refer suspects to the CI element for interrogation and further exploitation.

7003. CORDON AND SEARCH OPERATIONS

General

The timely seizure and exploitation of CI targets require a detailed and coordinated plan that has been prepared well in advance. CI elements, in most instances, cannot neutralize, guard or physically control targets without assistance. In some cases, this assistance must come from ground combat units for the seizure and protection of well-defended targets. In other cases, the assistance may be provided by combat support, combat service support, aviation units or even host country elements. It is essential that the required assistance be provided for during the planning phase.

The senior tactical unit commander will be the individual responsible for the conduct of the cordon and search operation. That commander will plan, with advice from CI, interrogation, CA and PSYOP personnel, the cordon that is usually deployed at night, and the search that normally begins at first light.

MAGTF CI personnel normally accompany the troops used in cordon and search operations to advise, assist, and examine and/or exploit the target at the earliest possible time. In some instances, it may be advantageous for CI personnel to rendezvous with the assigned troops at the target area. Except in unusual cases, the tactical effort takes precedence over the neutralization and exploitation of CI targets. If assistance in target seizure is not available, CI elements may have to rely on their own assets to neutralize or exploit targets. In friendly controlled areas, CI elements may coordinate through the JTF TFCICA to receive assistance from other JTF and services' CI elements, civil police, and security agencies.

The basic operation is the community cordon and search operation shown in figure 7-2.

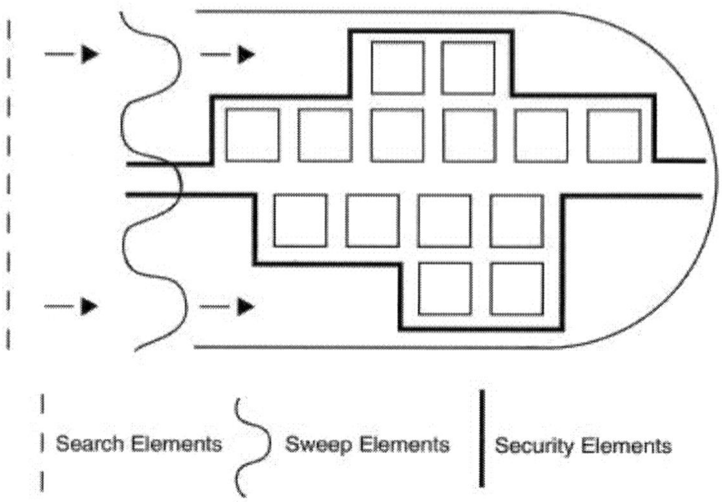

Figure 7-2. Example of a Community Cordon and Search Operation

Types and Conduct of Cordon and Search Operations

Community Operations

As the screening element sets up the collection or screening station (see figure 7-3), the sweep element escorts the residents toward the station, leaving behind one resident to care for family belongings, if required by law.

The search element follows behind the sweep element searching houses, storage areas, cemeteries etc., with dogs and metal detection equipment. CI personnel are searching for evidence of enemy intelligence collection operations including communications codes or other such paraphernalia. Each search element should include a CI team with an IT element as required, which will have a list of persons of CI interest.

In the collection or screening station, bring the residents to the collection area (or holding area) and then systematically lead them to specific screening stations. Enroute to the screening station, search each individual for weapons. Then lead the residents past the mayor or community leaders (enemy defectors or cooperating prisoners who will be hidden from view so they can uncompromisingly identify any recognizable enemy). These informants will be provided with the means to notify a nearby guard or a screener if they spot an enemy member. Immediately segregate this individual and interrogate by appropriate personnel.

At specific screening stations, ask the residents for identification, check against personalities list (black list), and search for incriminating evidence by electronic equipment.

Move suspected persons on for photographing, further interrogation or put them in the screening area detention point to be taken back to a base area or interrogation facility for detailed interrogation on completion of the operation.

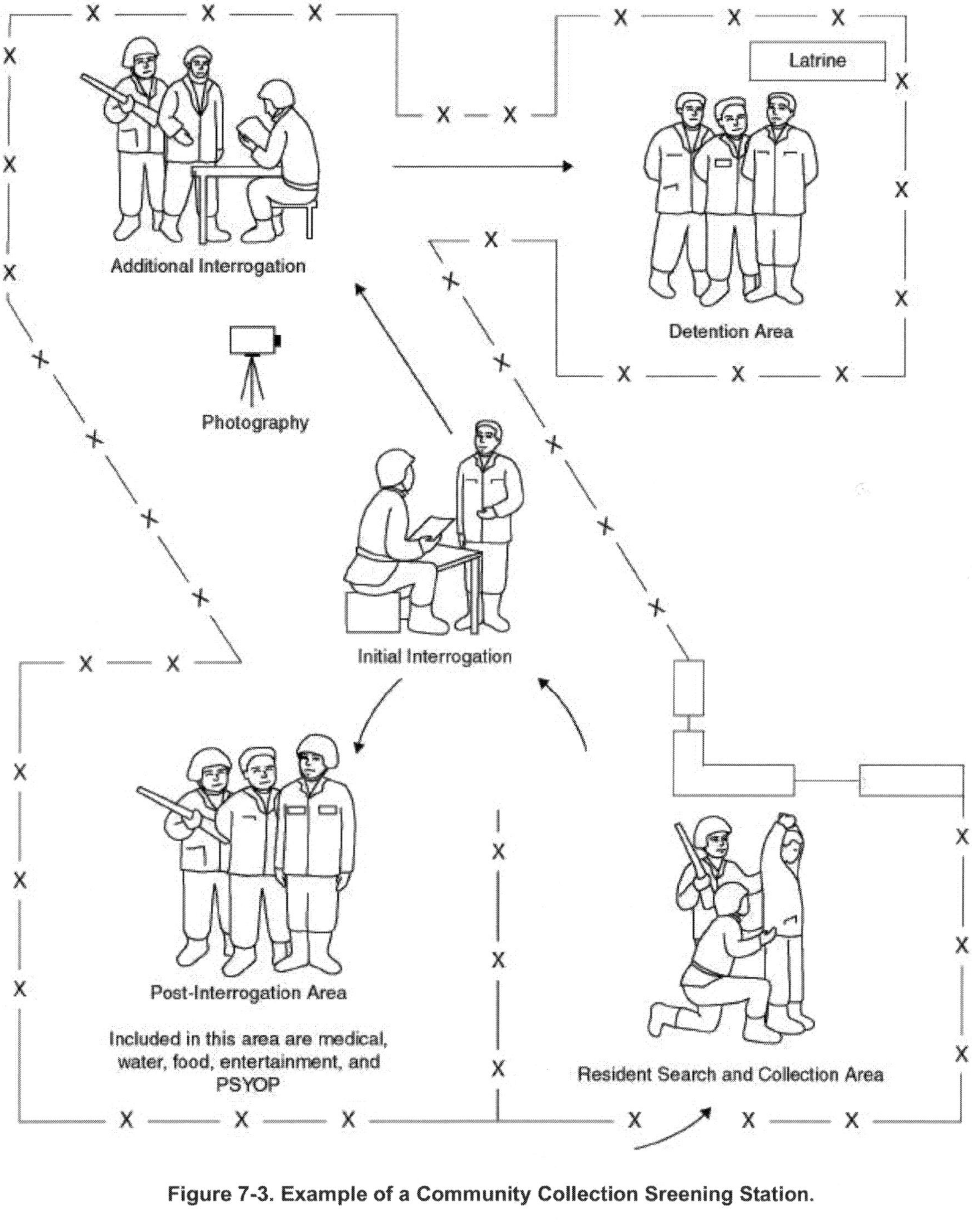

Figure 7-3. Example of a Community Collection Sreening Station.

Pass innocent residents through to the post screening area where they are provided medical assistance and other civic assistance, as well as entertainment and friendly propaganda.

Return any persons caught attempting to escape or break through the cordon immediately to the detention area.

When the operation is terminated, allow innocent individuals to return to their homes and remove the enemy suspects under guard for further interrogation. Photograph members of the community for compilation of a village packet, which will be used in future operations.

The second type of cordon and search operation is very frequently referred to as the soft or area cordon and search. This operation includes the cordoning and searching of a rather vast area (for example, a village area incorporating a number of hamlets, boroughs, town or villages that are subdivisions of a political area beneath country level).

Soft or area operation

This type of operation requires a larger military force to cordon off the area; a pooling of all paramilitary, police, CA, CI, and intelligence resources to conduct search and screening; and a formidable logistical backup. This kind of operation extends over a period of days and may take as long as a week or possibly longer.

While screening and search teams systematically go from community to community and screen residents, military forces sweep the area outside the communities over and over again to seek out anyone avoiding screening. As residents are screened, CI personnel will issue documents testifying to the fact that they were screened and if necessary, allow them restricted travel within the area.

Other population and resource control measures are used as well. Such an opportunity may allow the chance to issue new ID cards and photograph the area's residents.

As each community screening proceeds, send individuals who were designated for further interrogation to a centralized interrogation center in the cordoned area. Here, CI personnel will work with IT personnel and indigenous, police, and other security service interrogators.

Besides field files and other expedient facilities, a quick reaction force should be located at the interrogation center to react immediately to intelligence developed during the interrogations and from informants planted among the detainees.

7004. COUNTERINTELLIGENCE FORCE PROTECTION SOURCE OPERATIONS

CFSO's are flexible and aggressive collection operations conducted by CI personnel to quickly respond to the needs of the supported command. CFSO are focused on the collection of force protection information designed to assess threats from foreign intelligence collectors; provide early warning of impending attack; warn of sabotage or subversive activity against U.S. forces; identify and neutralize potential enemy infiltrations; provide information on local security forces; identify population and resource control measures; locate hostile or insurgent arms caches and safe havens; and identify local insurgent support personnel in regions where local security forces cannot or will not support U.S. operations. Additional policy guidance and procedures for the conduct of CFSOs in support of MAGTF operations is contained in classified MCO 003850.2, *Marine Corps Counterintelligence Force Protection Source Operations, (U)*. (See appendix D for the format of a CFSO Concept Proposal).

7005. TACTICAL CI INTERROGATION

Within the AO, there may be numerous people who are viewed as threats to security based solely on their presence in the combat zone. The number of suspect personnel varies. Frequently, it precludes detailed interrogation of all but a selected few that are of primary interest. CI personnel are partly dependent on such agencies as the provost marshal, civil affairs units, and IT platoons to identify suspect persons or persons of CI interest. In some situations, the number of persons volunteering information to CI operations permit concentration on those of the greatest potential interest or value. Most suspects are apprehended while trying to enter the area when their cover stories, which will closely parallel their true places of origin and identities, are exposed.

The CI or interrogation personnel's success in such interrogations is primarily dependent on questioning skill, linguistic ability or support, knowledge of the AO and adjacent areas, and familiarity with the intellectual, cultural, and psychological peculiarities of the persons encountered.

Types of Subjects

As the battle lines in combat change, entire segments of the population may be overrun. The local population in any area may also be increased by refugees and displaced persons (persons from other countries conscripted by enemy forces for labor). The following categories of persons are of CI interest:

- Refugees and displaced persons.
- Line crossers.
- Deserters from enemy units.
- Enemy intelligence personnel.
- Inmates of enemy detention camps.
- Members of underground resistance organizations seeking to join friendly forces.
- Collaborators with the enemy.
- Target personalities, such as black, gray or white list personalities.
- Volunteer informants.
- Persons who must be questioned because they are under consideration for employment with MAGTF units or for appointment as civil officials.

Objectives of CI Interrogators

CI interrogation in combat areas assists in the accomplishment of three major objectives:

- In the screening process, refugees whose very presence threatens overall security are removed from the battlefield.
- In detailed interrogations, enemy agents with espionage, sabotage, terrorist, or subversive missions are detected.
- The wide range of CI activities and types of interrogations permit the collection of information of value to other intelligence and security

agencies and to the planners of military operations. CI interrogators must be especially alert to obtain and report information of immediate tactical value, which may not have been previously obtained or reported.

Indicators Warranting Suspicion

CI personnel must be alert during interrogations for indications of intelligence activity. The following are indicators that, separately or collectively, may generate suspicion that a subject is in the employ of or acting in sympathy with enemy forces.

Access to Information or Targets

The interrogation should establish a subject's accessibility to potential targets, including the location at the time of apprehension.

A prospective terrorist, subversive, espionage or sabotage agent must have access to the information desired by the enemy or to the target installation to be destroyed to carry out the mission.

Technical Skills

The subject who has a mastery of one or several foreign languages and knowledge of radio operation or cryptography is questioned carefully on the nature and purpose of training in those fields.

Proficiency in certain technical skills is frequently an attribute of an espionage or sabotage agent. The subject's practical experience and work in those fields, during or shortly prior to the war, should give CI personnel cause for strong suspicion. The individual's story then must be closely examined.

Documents and Funds

An overabundance of documents and new documents of questionable authenticity are reason for doubt. They provide the basis for detailed questioning. Discrepancies in the document's contents or conflicts between data and the subject's story may lead to the detection of hostile agents. Unexplainable possession of large amounts of money, valuable jewelry or other items of great value are investigated carefully.

Pro-Enemy Background

Residence or travel in enemy territory, membership in a hostile party or known former collaboration with the enemy are facts of obvious importance. CI personnel must determine whether the subject is actually in sympathy with the enemy or has acted merely to serve the subject's best interests with regard to life, welfare of family or property.

Family in Enemy-Held Territory

Enemy pressure is often applied to individuals whose families reside under enemy control. Individuals who have family members threatened with death/torture/incarceration by the enemy will always be a threat to friendly forces.

Inconsistent Story

Small discrepancies in the subject's story may be important. Contradictions in a subject's story do not warrant jumping to conclusions. However, CI personnel must remain alert to all possibilities. Allowances must be made for defective memory or lack of logic due to emotional stress.

The following discrepancies may be warning signals to the CI interrogator:

- Distance compared to travel time.
- Accent peculiar to an area the subjects refuse to acknowledge as their own.
- Unreasonable explanation of deferment, exemption or discharge from military service.
- Exemption from labor conscription.
- Implausible reasons for risking the crossing of combat lines.

Suspicious Actions or Activities

Indigenous persons displaying unusual interest in troop units and equipment or loitering persistently in the vicinity of troop units and installations, without reasonable explanation, are sufficient to warrant interrogation for the purpose of clarifying the status of persons so involved.

Violations of Civil or Military Regulations. Mere violation of military regulations in an area controlled by the military may be relatively unimportant to CI elements. These violations may be mandatory registration, curfews, travel restrictions or declaration of weapons. However, the motives which cause such violations despite severe penalties may be compelling and possibly of great interest to CI personnel.

Modus Operandi (MO). Frequent similarity in tactics of hostile agents working for the same enemy agency, their means of contact with their agent handlers, type of cover story, and manner of collecting and reporting their information may lead to identification of suspects with a known enemy agency or group. Established patterns of activity or behavior of enemy agents are disseminated to other intelligence and security agencies to assist in the identification of agents still operating.

Screening or Initial Interrogation

Initial interrogation and screening are generally synonymous. However, initial interrogation indicates that there will be a follow-up, detailed interrogation while screening involves the selection, by brief questioning, of a relatively small number of persons from a large group for detailed interrogation. In either case, the technique, purpose, and scope of the questioning are generally the same. The object is to select, for detailed interrogation, a reasonable number of persons who appear to be knowledgeable of matters of CI interest. Initial interrogation or screening is generally concerned with identity, background, recent activities, travel or escape routes, and information of immediate value. Documents and personal belongings of a subject are examined. Then the circumstances of apprehension are studied. Finally, the available files are checked.

Detailed Interrogation

Detailed CI interrogations may be conducted in joint interrogation facilities or at MAGTF interrogation sites/collection points established by G-1/S-1s and manned by interrogators and CI personnel. Detailed interrogation does not differ radically from the initial interrogation except that attention is now focused on individuals who are suspect or who are known to have extensive information of interest. A study of the initial interrogation report, examination of the subject's documents and belongings, and checks of available files and information must be conducted and analyses made in preparation for the interrogation.

Details of the subject's personal history must be reviewed. Should the subject admit being an enemy agent, the individual becomes an important source of information on enemy intelligence methods of operation and, perhaps, on identities of other hostile agents. This leads to exhaustive interrogations on such issues as hostile intelligence training and missions assigned. However, CI personnel must be alert to the possible insertion of confusion agents.

The suspect, or any person being interrogated, may also be an important source of information of valuable strategic and/or tactical intelligence.

Questioning usually follows a logical sequence to avoid confusing the subject and to facilitate reporting. However, an illogical sequence may be used to purposely to confuse the subject to inadvertently contradict him. The interrogator must be alert for discrepancies and retain psychological advantage.

See the employment paragraph for additional information on CI interrogations.

7006. CI INVESTIGATIONS

CI investigations are conducted when sabotage, espionage, treason, sedition or subversive activity is suspected or alleged. CI investigations may also be conducted regarding security matters and defections of friendly personnel to the enemy. The primary purpose of each investigation is to identify, neutralize, and exploit information of such a nature, form, and reliability, that may determine the extent and nature of action necessary to counteract the threat and enhance security. The investigation is a duly authorized, systematic, detailed examination/inquiry to uncover and report the facts of a matter. While facts, hearsay, information, opinions, allegations, and investigators' comments may make a significant contribution, they should be clearly labeled as such in the report of investigation. Investigations are generally incident investigations concerning acts or activities committed by, or involve, known or unknown persons or groups of persons. CI agents conducting investigations must have a thorough understanding of the objectives and operations of foreign espionage, sabotage, and subversive organizations.

The Naval Criminal Investigative Service Manual, NIS-3, Manual for Investigations, may be used as a guide for investigation techniques and procedures..

Conduct of CI Investigations

CI investigations use basic investigative techniques and procedures. The primary purpose of the investigation is to provide the commander with sufficient factual information to reach a decision or ensure security of the command. Investigations may be conducted overtly or discreetly depending on the type of investigation and the AO. Investigations will normally include the examination of records, interviews or interrogations, and evidence. Surveillance and the conduct of raids and searches may also be appropriate as the investigation progresses.

On assuming primary CI jurisdiction, MAGTF CI investigations are conducted in accordance with guidance and instructions published by a higher authority.

CI personnel are normally responsible for conducting security investigations of indigenous personnel employed by MAGTF elements. They may also be involved in the investigation of indigenous personnel retained in official civilian positions.

Certain unique problems are involved in conducting investigations of indigenous personnel in a tactical environment. One problem that may hinder investigation is lack of files and records in the repositories of civilian police and investigative agencies. This documentation may have been destroyed or removed during tactical operations. Every effort must be made to check files and records that are available.

Another problem may be lack of qualified personnel to perform investigations. Special investigative techniques, such as the use of polygraph examinations by criminal investigative personnel, may be required.

A problem that presents a security threat to the MAGTF is the use of indigenous personnel for a wide range of support functions. Indigenous civilian employees may be sympathetic to the enemy's cause or coerced to serve the enemy cause. If using indigenous personnel, caution must be exercised to preclude the enemy's collection of useful information, both classified and unclassified. In addition to the initial security investigation, continual checks are made on indigenous employees. CI elements maintain close liaison with civil affairs units responsible for providing civilian labor to the MAGTF.

Investigative Plan

In the conduct of a CI investigation if evidence or indicators of criminal activity are discovered, this information should be provided to the CID of the Provost Marshals Office. Information should only be provided if such disclosure does not compromise the ongoing CI investigation.

When required, CI personnel formulate an investigative plan at each command level down to and including the individual CI Marine. Normally, the lead investigative element will develop the plan. The investigative plan must be updated as new developments arise, including an ongoing analysis of the results. Although this list is not all encompassing, an investigative plan should include as many of the following planning considerations as applicable:

- Purpose of the investigation.
- Definition of the problem.
- Phases or elements of the investigation that have been assigned.
- Whether the investigation is to be conducted overtly or discreetly.
- Priority and time permitted for completion.
- Special instructions or restrictions.
- Information from the unit or office files.

- Methods and sources used including surveillance and polygraph support.
- Coordination required.

Order of Investigation

CI investigations vary and investigative plans will be different. The following actions are typically conducted during an investigation. Tailor investigative plans to each investigation. Investigative actions selected should be sequenced to ensure a swift and successful completion of the investigation.

- Files and records check for pertinent information.
- Individual interviews for additional information and leads.
- Exploitation of new leads and consolidation of available data for analysis and planning a course of action.
- Surveillance, both physical and technical, of the subject(s) to be investigated.
- Interrogation or interview of the subject(s) to prove or disprove the allegations.
- Polygraph examination.

Investigative Techniques

CI personnel use the following basic techniques in CI investigations and operations, as appropriate:

- Examine records to locate, gain access to, and extract pertinent data from diverse official and unofficial documents and records.
- Conduct interviews to obtain information. The type of interviews conducted depends on the investigation.
- Use interrogation and elicitation techniques as additional methods to gather information.
- Conduct physical and technical surveillance to augment other investigative activities.
- Conduct cordon, search, and seizure when necessary. Do not conduct searches unless directed by proper authority. CI personnel may coordinate this activity with law enforcement agencies, depending on the nature of the investigation.

Files and Records

Checking files and records for pertinent information on the subject of the investigation is the first action in CI investigations. Checks should begin with local unit files and expand to include other possible sources. The full exploitation of record examination as an investigative tool depends on several factors that CI personnel must consider.

CI personnel must know what, where, by whom, and for what purpose records are maintained throughout the AO. Upon assignment to an operational unit, the initial orientation should stress that CI personnel are thoroughly familiar with records of assistance in investigations.

Most records are available to CI personnel upon official request. If all efforts to obtain the desired information through official channels are unsuccessful, the information or records cannot be subpoenaed unless legal proceedings are initiated.

There are occasions when documentary information or evidence is best obtained through other investigative means. The possibility of intentional deception or false information in both official and unofficial records must be considered. Because data is recorded in some documentary form does not ensure reliability. Many recorded statistics are untrue or incorrect, particularly items of biographical data. They are often repetitious or unsubstantiated information provided by the subject being investigated and not to be confused with fact.

Reliability of records varies considerably according to the area and the status of the agency or organization keeping the records. Records found in highly industrialized areas, for example, are more extensive and generally far more reliable than those found in underdeveloped areas. Until experience with a certain type of record has been sufficient to make a thorough evaluation, treat the information with skepticism.

In CI investigations, the absence of a record is often just as important as its existence. This is especially important in the investigation of biographical data furnished by the subject being investigated. The systematic and meticulous examination of records to confirm or refute the subject's story is very often the best means of breaking the cover story of an enemy intelligence agent.

The types and content of records vary markedly with the AO. Regardless of the area, CI personnel must be aware of the types of records to be used in conducting investigations. Available records include police and security agencies, allied agencies, vital statistics, residence registration, education, employment, citizenship, travel, military service, foreign military records, finance records, and organization affiliation.

Police and Security Agencies

Records of value are often found at local, regional, and national police agencies. Most nations maintain extensive personality files covering criminals, CI investigative subjects, victims, and other persons who have come to official police attention because of actual or alleged criminal activity. Police and security agency files are usually divided into subcategories. CI personnel must be familiar with the records system to ensure pertinent files actually have been checked.

Police interest in precise descriptive details, including photographs and fingerprint cards, often make police records particularly valuable and usually more reliable than comparable records of other agencies.

Allied Agencies

Access to records of allied intelligence agencies often depends on the personal relationship between U.S. CI personnel and the custodian of the records of interest. Such examinations are normally the assigned responsibility of a CI liaison officer. Liaison also may be necessary with other agencies when the volume of records examinations dictate the need for a single representative of the CI element. It may be necessary, due to the sensitivity of a particular investigation, to conceal specific interest in a person whose name is to be checked. Here, the name of the individual may be submitted routinely in the midst of a lengthy list of persons (maybe five to seven) who are to be checked.

Vital Statistics

The recording of births, deaths, and marriages is mandatory in nearly every nation, either by national or local law. In newly developed countries, however, this information may be maintained only in family journals, bibles, or in old records. Confirmation of such dates may be important. Records sought may be filed at the local level, as is usually the case in overseas areas; or they may be kept at the state or regional level, such as with state bureaus of vital statistics in the U.S. Rarely will original vital statistics records on individuals be maintained centrally with a national agency.

Residence Registration

Some form of official residency registration is required in most nations of the world. The residence record may be for tax purposes and probably will be found on file at some local fiscal or treasury office. When the residence record is needed for police and security purposes, it is usually kept in a separate police file. Residence directories, telephone books, and utility company records also may be used.

Education

Both public and private schools from primary grades through universities, have records that can serve to verify background information. The school yearbook or comparable publication at most schools usually contains a photograph and brief resume of the activities of each graduating class member. These books are a valuable record for verification and as an aid to locating leads. Registrar records normally contain a limited amount of biographical data but a detailed account of academic activities.

Employment

Personnel records usually contain information on dates of employment, positions held, salary, efficiency, reason for leaving, attendance record, special skills, and biographical and identifying data. Access to these records for CI personnel are relatively simple in the U.S., but may prove difficult in some overseas areas. In such areas, it may be possible to obtain the records through liaison with local civil authorities or through private credit and business rating firms. Depending on the AO, there may be local, regional, or national unemployment and social security program offices. Records of these offices often contain extensive background material. Usually, these data represent unsubstantiated information provided by the applicant and cannot be regarded as confirmation of other data obtained from the same individual.

Refugee records (particularly those of private welfare groups) are used as a source of leads rather than for verification of factual data, since they have been found to be unreliable in nearly all AOs.

Citizenship

Immigration, nationalization, passport, and similar records of all nations contain data regarding citizenship status. In most instances, an investigation has been undertaken to verify background information contained in such records; therefore, these records are generally more reliable than other types. Records of both official and private refugee welfare and assistance agencies also provide extensive details relating to the citizenship status of persons of CI interest.

Travel

A system of access to records of international travel is especially important to overseas CI operations. Such records include customs records, passport and visa applications, passenger manifests of commercial carriers, currency exchange files, transient residence registrations, private and government travel agency records, and frontier control agency files.

Military Service

Records of current and past members of the armed services of most nations are detailed and usually accurate.

Foreign Military Records

Access to foreign military records in overseas areas may be difficult. In cases where it is not possible to examine official records, leads or pertinent information may be obtained from unofficial unit histories, commercially published documents, and files of various veterans' organizations. Special effort must be made to locate some form of record that confirms or denies an individual's service in a particular unit or the existence of the unit at the time and place the individual claims to have served. OOB and personality files of various intelligence services also may be helpful.

Since listing or claiming military service is a convenient means of accounting for periods of time spent in intelligence activities or periods of imprisonment, it is frequently a critical item in dealing with possible enemy agents.

Finance Records

Finance records are an important source of information. They may provide information to indicate whether a person is living beyond one's means. They may provide numerous leads such as leave periods and places, and identification of civilian financial institutions.

Organizations are often established as front groups or cover vehicles for foreign intelligence operations.

Organization Affiliation

Many organizations maintain records that may be of value to a particular investigation. Examples are labor unions; social, scientific, and sports groups; and cultural and subversive organizations. CI personnel should research these organizations. But when seeking sources of information, they must be thoroughly familiar with the organization before attempting to exploit it.

Interrogation Techniques

Interrogation is obtaining the maximum amount of usable information through formal and systematic questioning of an individual. CI interrogations should be conducted by at least two CI personnel.

CI personnel use interrogation techniques when encountering a hostile source or other subject being investigated. The self-preservation instinct is stimulated in an individual who is considered the subject. This deep-rooted reaction is frequently reflected in stubborn resistance to interrogation. The subject may consider the interrogation as a battle of wits where the subject has much to lose. The subject may look upon the CI interrogator as a prosecutor.

When interrogating a subject, CI personnel must keep in mind the two-fold objective of the interrogation:

- Detection and prevention of activity that threatens the security of the MAGTF.
- Collection of information of intelligence interest.

When preparing for an interrogation, CI personnel should—

- Gather and become completely familiar with all available material concerning the subject and the case.
- Be familiar with those legal principles and procedures that may apply to the case at hand. Legal requirements may differ depending on: whether the U.S. is at war or in a military occupation; status of force agreements; whether the subject being interrogated is a U.S. citizen or an EPW.
- Determine the best way to approach the subject. Previous investigative efforts may have determined that the subject is under great psychological pressure; therefore, a friendly approach might work best. CI personnel should carefully consider the approach and the succeeding tactics, to ensure that nothing the interrogator does will cause the subject to confess to a crime not committed.

Before an interrogation, CI personnel must ensure the following:

- The interrogation room is available and free of distractions.
- If recording equipment is used, it is installed and operationally checked.
- Participants in the interrogation team are thoroughly briefed on the case and interrogation plan.
- Sources or other persons to be used to confront the subject are available.
- Arrangements are made to minimize unplanned interruptions.
- As appropriate, arrangements are made for the subject to be held in custody or provided billeting accommodations.

When conducting the interrogation, the following points are important:

- Use background questioning to provide an opportunity to study the subject face-to-face.
- Avoid misinterpretation and impulsive conclusions. The fact that the person is suspected may create reactions of nervousness and emotion.
- Do not allow note taking to interfere with observing the subject's reaction.
- Seek out all details concerning the subject's implication in a prohibited activity.
- Examine each of the subject's statements for its plausibility, relationship to other statements or to known facts, and factual completeness. Discrepancies requiring adjustment frequently weaken the subject's position.
- Attempt to uncover flaws in details not considered relevant to the issue; finding the story's weakness is the key to a successful interrogation.
- Build up to a planned final appeal as a sustained and convincing attack on the subject's wall of resistance. Eloquent and persuasive reasoning and presenting the facts of the case may succeed where piecemeal

consideration of evidence failed to produce a confession. This appeal may be based on overwhelming evidence, on contradictions, story discrepancies, or the subject's emotional weaknesses.

- Obtain a sworn statement if the subject wants to confess. If the subject has been given an explanation of individual rights under Article 31, Uniform Code of Military Justice (UCMJ), or the 5th Amendment to the U.S. Constitution, any unsworn statement normally can be used in court. If the subject is neither a U.S. citizen nor a member of the armed forces, requirements will be stipulated in the unit's SOP.

Elicitation

CI personnel may use polygraph examinations as an aid to CI interrogations and investigations of intelligence operations, but only at the direction of higher headquarters.

Elicitation is gaining information through direct communication, where one or more of the involved parties is not aware of the specific purpose of the conversation. Elicitation is a planned, systematic process requiring careful preparation.

Preparation

Apply elicitation with a specific purpose in mind.

The objective, or information desired, is the key factor in determining the subject, the elicitor, and the setting.

Once the subject has been selected because of access or knowledge of the desired information, numerous areas of social and official dealings may provide the setting.

Before the approach, review available intelligence files and records, personality dossiers, and knowledge possessed by others who have previously dealt with the subject. This will help to determine the subject's background, motivation, emotions, and psychological nature.

Approach

There are two basic elicitation approaches: flattery and provocation.

Approach the subject in normal surroundings to avoid suspicion. The following variations to these approaches may be used:

- By appealing to the ego, self-esteem or prominence of the subject, you may be able to guide the individual in a conversation on the AO.
- By soliciting the subject's opinion and by insinuating that the subject is an authority on a particular topic, you may be able to obtain desired information.
- By adopting an unbelieving attitude, you may be able to cause the subject to explain in detail or to answer out of irritation. CI personnel should not provoke the subject to the point where rapport is broken.
- By inserting bits of factual information on a particular topic, you may be able to influence the subject to confirm and further expound on the topic. Use this approach carefully since it does not lend itself to sudden impulse. Careless or over use of this technique may give away more information than gained.
- By offering sincere and valid assistance, you may be able to determine the subject's specific AOI.

Conversation

Once the approach has succeeded in opening the conversation, devise techniques to channel the conversation to the AOI. Some common techniques include:

- An attempt to obtain more information by a vague, incomplete, or a general response.
- A request for additional information where the subject's response is unclear; for example, "I agree; however, what did you mean by ________?"
- A hypothetical situation that can be associated with a thought or idea expressed by the subject. Many people who would make no comment concerning an actual situation will express an opinion on hypothetical situations.

Sabotage Investigations

Sabotage is defined as an act, the intent to damage the national defense structure. Intent in the sabotage statute means knowing that the result is practically certain to follow, regardless of any desire, purpose, or motive to achieve the result. Because the first indication of sabotage normally will be the discovery of the injury, destruction, or defective production, most sabotage investigations involve an unknown person or persons. We expect acts of sabotage, both in overseas AOs and in CONUS, to increase significantly in wartime. Sabotage is a particularly effective weapon of guerrilla and partisan groups, operating against logistic and communications installations in occupied hostile areas, and during insurgencies. Trained saboteurs sponsored by hostile guerrilla, insurgent, or intelligence organizations may commit acts of sabotage. Individuals operating independently and motivated by revenge, hate, spite, or greed may also conduct sabotage. In internal defense or limited war situations where guerrilla forces are active, we must be careful to distinguish among those acts involving clandestine enemy agents, armed enemy units, or dissatisfied friendly personnel. Normally, we categorize sabotage or suspected sabotage according to the means employed. The traditional types of sabotage are incendiary, explosive, and mechanical. In the future, nuclear and radiological, biological, chemical, magnetic, and electromagnetic means of sabotage will pose an even greater threat to military operations. We must preserve and analyze the incident scene before evidence is altered or destroyed.

Sabotage investigations require immediate action. The possibility exists that the saboteur may still be near the scene, or that other military targets may require immediate or additional security protection to avoid or limit further damage.

Questions

The investigation must proceed with objective and logical thoroughness. The standard investigative interrogatives apply.

- Who—determine a list of probable suspects and establish a list of persons who witnessed or know about the act.
- What—determine what military target was sabotaged and the degree of damage to the target (both monetary and operational).
- When—establish the exact time when the act of sabotage was initiated and when it was discovered; confirm from as many sources as possible.

- Where—determine the precise location of the target and its relation to surrounding activities.
- Why—establish possible reasons for the sabotage act through the investigation of suspects determined to have had motive, ability, and opportunity to accomplish the act.
- How—establish the type of sabotage (such as incendiary, explosive, chemical) and determine the procedures and materials employed through investigation and technical examination and analysis.

Investigative Actions

An outline of possible investigative actions used to investigate alleged or suspected sabotage incidents follows.

- Obtain and analyze the details surrounding the initial reporting of the incident. Establish the identity of the person reporting the incident and the reasons for doing so. Determine the facts connected with the reported discovery of the sabotage and examine them for possible discrepancies.
- Examine the incident scene as quickly as possible. CI personnel must attempt to reach the scene before possible sources have dispersed and evidence has been disturbed. They will help MP personnel protect the scene from disruption. The MPs will remove all unauthorized persons from the area, rope off the area as necessary, and post guards to deny entrance and prevent anything from being removed. Although CI personnel should help MP investigators at the sabotage scene, they should not interfere with the crime scene investigation.
- Preserve the incident scene by taking notes, making detailed sketches, and taking pictures. Arrange for technical experts to help search the scene and collect and preserve physical evidence and obtain all possible clues. Arson specialists, explosive experts, or other types of technicians may be required. Take steps to prevent further damage to the target and to safeguard classified information or material. (d) Interview sources and obtain sworn statements as soon as possible to reduce the possibility of forgetting details or comparing stories.
- Determine the necessary files to be checked. These will be based on examination of the incident scene and by source interviews. CI conducts such action only with the MAGTF PM, retaining sabotage scene expertise and responsibility.

Files checks should include background information on sources and the person or persons who discovered or reported the sabotage.

Files of particular importance may include—

- Friendly unit MO files.
- Partisan, guerrilla, or insurgent activity files.
- Local police files on arsonists.
- Local police MO files.
- Host country's intelligence agency MO files.
- Terrorist modus operandi files.
- Provost marshal files.

Study all available information such as evidence, technical and laboratory reports, statements of sources, and information from informants in preparation for interrogation of suspects.

CI Walk-In Interviews

A walk-in is defined as an individual who seeks out MAGTF authorities to volunteer information believed to be of intelligence value. The primary concern of CI personnel is to obtain all information of intelligence and CI value. They must be alert to detect whether the source provides leads for further exploitation.

Motivation

When interviewing such persons, CI personnel must consider the source's motives for divulging information. The motivation may not be known, and sources may not be truthful about their motives. If the motive can be determined early in the interview, however, it can be valuable in evaluating the information supplied and in determining the nature and extent of the source's knowledge and credibility. Motivation includes, but is not limited to: ideology, personal gain, protection of self or family ties, fear, misunderstanding of the function and mission of the MAGTF, mental instability, and revenge.

Preparation

In preparing and conducting a walk-in interview, CI personnel—

- Should adapt to the intellectual level of the source, exercise discretion, and avoid controversial discussions.
- Obtain names and whereabouts of other individuals who may directly or indirectly know the same information.
- Remember security regulations and make no commitments that cannot be fulfilled.

Conduct of a Walk-in Interview

Put the source at ease. After determining that a walk-in source has information of intelligence value, display the appropriate credentials.

- **Take the source to a private place to conduct the interview.** The initial attitude frequently affects the success of the interview. The atmosphere should be pleasant and courteous, but professional. In accordance with the Privacy Act of 1974, if the source is a U.S. citizen or alien lawfully authorized permanent residence status, the source must be given a four point Privacy Act Advisement to include authority, principal purpose, routine uses, and voluntary and mandatory disclosure, prior to obtaining personal information. Ask for some form of identification, preferably one with a picture.
- **Record the pertinent data from the ID card and tactfully exit the room.**
- **Check the office source or informant files to see what information on the source is on file using the identity information just obtained from the source.** Determine if the source is listed as a crank, has a criminal record or has reported information in the past, and if so, what was the validity and value of that information.
- **Continue with the interview if the source is listed as a crank or a nuisance but include this information in the appropriate memorandum.**

- **Let the source tell the story.** Suggest that the source start the story from the beginning, using the source's own words. Once started, let the source talk without interruption. CI personnel should, however, guide the source back when straying from the basic story. From time to time, interject a word of acknowledgment or encouragement. At no time should CI personnel give any indication of suspicion or disbelief, regardless of how incredulous the story may seem. While the source gives an account for the first time, take minimal notes. Taking notes could distract the source or the CI interviewer. Instead, pay close attention and make mental notes of the salient points as a guide for subsequent detailed interviewing.
- **Review the story with the source and take notes.** Once the source has finished telling the basic story, the answer will generally be freely given to specific questions on the details. Being assured that the information will be kept in strict confidence, the source will be less apprehensive of your note taking. Start at the beginning and proceed in a chronological order, using the salient features of the source's account. Interview the source concerning each detail in the account so that accurate, pertinent information is obtained, meticulously recorded, and that the basic interrogatives are answered for every situation. This step is crucial.
- **Develop secondary information.** The story and background frequently indicate that the source may have further information of significant intelligence interest. Also develop this information fully.
- **Terminate the interview.** When certain that the source has no further information, close the interview in a manner that leaves a favorable impression. At this point in the interview, ask the source, point blank, what was the motivation to come in and report the information, even if the source volunteered a reason earlier in the interview. Obtain a sworn statement from the source, regarding the information, if appropriate. It is best to have the source write (or type) the statement. Ask for full name, rank or occupation, duty position, unit of assignment, social security number, date and place of birth, type of security clearance and level of access, and full current address. Determine who else knows about the incident or situation, either directly or indirectly. Determine the source's desires regarding release of identity. Determine the source's willingness to be recontacted by CI personnel or those from another agency should the need arise regarding the information provided. Obtain recontact information from the source (work or residence). If the source is a U.S. citizen or alien lawfully authorized permanent residence status, have a disclosure warning executed and the affirmation attached to the report as an exhibit. Finally, express appreciation for the information received.

7007. CAPTURED MATERIAL EXPLOITATION

If the situation permits, CI personnel should exploit enemy installations immediately following neutralization or capture.

An installation is searched thoroughly for documents, equipment, and other material of intelligence or CI interest, which will be marked and rapidly transported to MAGTF IT. In some instances, it may be desirable to retain the documents or material within the installation for thorough examination by technical intelligence personnel or other specialists. Due to the risks of booby traps, mines, and explosives, extreme caution must be used when searching installations known or suspected to have been occupied by the enemy.

Figure 7-4 is an example of a captive/document/equipment tag.

Do not remove tag from captive/document/equipment	
Captive Tag	Instructions (Captive Tag)
Tag number____________	1. Complete upper half of tag for each captive.
Date/time of capture____________	2. If captive has document, check yes. Complete and detach lower half of tag.
Place of capture	3. Securely affix tag to captive.
(coord.)____________	Additional information: ____________
Circumstances of capture__________	____________
Weapons___No ___Yes ___Type	____________
Document ___No ___Yes (if yes complete lower half of tag)	____________
Capturing unit____________	4. Additional Information: __________
____________	____________
Do not remove tag from captive/document/equipment.	

Figure 7-4. Captive/Document/Equipment Tag.

7008. CI TECHNICAL COLLECTION AND INVESTIGATIVE TECHNIQUES

CI investigators use a variety of technical investigative techniques, of which the following are those most typically employed in support of MAGTF operations: TSCM; electronic surveillance; investigative photography and videotaping; and polygraphs.

Technical collection and investigative techniques can contribute materially to the overall CI investigation and activities. They can assist in supplying the commander with factual information to base decisions concerning the security of the command.

Technical Surveillance Countermeasures

TSCM versus TEMPEST

TSCM is a defensive CI measure used in counterespionage activities. It is concerned with all signals leaving a sensitive or secure area, including audio, video, digital or computer signals. There is a definite distinction between TSCM and TEMPEST.

TEMPEST is the unintentional emanation of electronic signals outside a particular piece of equipment. Information systems, computers, and electric typewriters create such signals. The words to focus on in TEMPEST are known and unintentional emanations. TEMPEST is controlled by careful engineering or shielding.

TSCM is concerned with the intentional effort to gather intelligence by foreign intelligence activities by emplacing covert or clandestine devices into a U.S. facility, or modifying existing equipment within that area. Mostly, intelligence gained through the use of technical surveillance means will be accurate, as people are unaware they are being monitored. At the same time, the implanting of such technical surveillance devices is usually a last resort.

Threat

Enemy intelligence and security forces, their agents, and other persons use all available means to collect sensitive information. One way they do this is by using technical surveillance devices, commonly referred to as bugs and taps. Such devices have been found in U.S. facilities worldwide. Security weaknesses in electronic equipment used in everyday work have also been found worldwide. The enemy easily exploits these weaknesses to collect sensitive or classified conversations as well as the information being processed. They are interested in those things said in (supposed) confidence, since they are likely to reveal future intentions. It should be stressed that the threat is not just audio, but video camera signals, as well as data. Devices are usually placed to make their detection almost impossible without specialized equipment and trained individuals.

The TSCM Program

The purpose of the TSCM program is to locate and neutralize technical surveillance devices that have been targeted against U.S. sensitive or secure areas. The TSCM program identifies and enables the correction of exploitable technical and physical security vulnerabilities. The secondary, and closely interrelated purpose, provides commanders with a comprehensive evaluation of their facilities' technical and physical security postures.

DODINST 5240.5, *DOD Technical Surveillance Countermeasures Survey Program*, and OPNAVINST C5500.46, *Technical Surveillance Countermeasures*, govern the implementation of this program.

The TSCM program includes four separate functions; each with a direct bearing on the program.

- **Detection.** Realizing that the threat is there, the first and foremost function of the TSCM program is to detect these devices. Many times these devices cannot be easily detected. Occasionally, TSCM personnel will discover such a device by accident. When they discover a device, they must neutralize it.
- **Nullification.** Nullification includes both passive and active measures used to neutralize or negate devices that are found. An example of passive nullification is soundproofing. But soundproofing that covers only part of a room is not helpful. Excessive wires must be removed, as they could be used as a transmission path from the room. Nullification also refers to those steps taken to make the emplacement of technical surveillance systems as difficult as possible. An example of active nullification is the removal of a device from the area.
- **Isolation.** The third function of the TSCM program is isolation. This refers to limiting the number of sensitive or secure areas and ensuring the proper construction of these areas.
- **Education.** Individuals must be aware of the foreign intelligence threat and what part they play should a technical surveillance device be detected. Additionally, people need to be alert to what is going on in and around their area, particularly during construction, renovations, and installation of new equipment.

The TSCM program consists of CI technical investigations and services (such as surveys, inspections, pre-construction advice and assistance) and technical security threat briefings. TSCM investigations and services are highly specialized CI investigations and are not to be confused with other compliance-oriented or administrative services conducted to determine a facility's implementation of various security directives.

- **TSCM Survey.** This is an all-encompassing investigation. This investigation is a complete electronic, physical, and visual examination to detect clandestine surveillance systems. A by-product of this investigation is the identification of physical and technical security weaknesses that could be exploited by enemy intelligence forces.
- **TSCM Inspection.** Normally, once a TSCM survey has been conducted, it will not be repeated. If TSCM personnel note several technical and physical weaknesses during the survey, they may request and schedule an inspection at a later date. In addition, they will schedule an inspection if there has been an increased threat posed to the facility or if there is some indication that a technical penetration has occurred in the area. No facility, however, will qualify automatically for recurrent TSCM support.
- **TSCM Pre-construction Assistance.** As with other technical areas, it is much less expensive and more effective to build in good security from the initial stages of a new project. Thus, pre-construction assistance is designed to help security and construction personnel with the specific requirements needed to ensure that a building or room will be secure and built to standards. This saves money by precluding costly changes later on.

Request for TSCM Support

Requests for, or references to, a TSCM investigation will be classified SECRET, marked with the protective security marking, and receive limited dissemination (to include no dissemination to any non-U.S. recipient). The fact that support is scheduled, in progress, or completed, is classified SECRET.

No request for TSCM support will be accepted via nonsecure means. Nonsecure telephonic discussion of TSCM support is prohibited.

Requests will be considered on a case-by-case basis and should be forwarded through the chain of command via the unit's intelligence officer or security manager.

When requesting or receiving support, the facility being inspected must be complete and operational, unless requesting pre-construction advice and assistance. If any additional equipment goes into the secure area after the investigation, the entire area is suspect and the investigation negated.

Fully justified requests of an emergency nature, or for new facilities, may be submitted at any time, but should be submitted at least 30 days before the date the support is required.

The compromise of a TSCM investigation or service is a serious security violation with potentially severe impact on national security. Do not compromise the investigation or service by any action that discloses to unauthorized persons that TSCM activity will be, is being, or has been conducted within a specific area.

Compromises

Unnecessary discussion of a TSCM investigation or service, particularly within the subject area, is especially dangerous. If a listening device is installed in the area, such discussion can alert persons who are conducting the surveillance and permit them to remove or deactivate their devices. When deactivated, such devices are extremely difficult to locate and may require implementation of destructive search techniques. In the event a

TSCM investigation or service is compromised, the TSCM team chief will terminate the investigation or service at once. Report the circumstances surrounding the compromise of the investigation or service to the supported unit or installation's intelligence officer or security manager.

Completion

When a TSCM survey or inspection is completed, the requester is usually given reasonable assurance that the surveyed area is free of active technical surveillance devices or hazards.

TSCM personnel inform the requester about all technical and physical security vulnerabilities with recommended regulatory corrective actions.

The requester should know that it is impossible to give positive assurance that there are no devices in the surveyed area.

The security afforded by the TSCM investigation will be nullified by the admission to the secured area of unescorted persons who lack the proper security clearance. The TSCM investigation will also be negated by—

- Failing to maintain continuous and effective surveillance and control of the serviced area.
- Allowing repairs or alterations by persons lacking the proper security clearance or not under the supervision of qualified personnel.
- Introducing new furnishings or equipment without a thorough inspection by qualified personnel.

Subsequent Security Compromises

Report immediately via secure means to the intelligence officer or security manager the discovery of an actual or suspected technical surveillance device. Information concerning the discovery will be handled at a minimum of SECRET. Installation or unit security managers will request an immediate investigation by the supporting CI unit or supporting TSCM element.

Electronic Surveillance

Electronic surveillance is the use of electronic devices to monitor conversations, activities, sound, or electronic impulses. It aids in conducting CI investigative activities. Various directives regulate the use of wiretapping and electronic eavesdropping and must be strictly adhered to by CI personnel.

Technical Surveillance Methodology

Technical surveillance methodology (including that employed by enemy intelligence and security forces) consists of—

Pickup Devices. A typical system involves a transducer (such as a microphone, video camera, or similar device) to pick up sound or video images and convert them to electrical impulses. Pickup devices are available in practically any size and form. They may appear to be common items, such as fountain pens, tie clasps, wristwatches, or household or office fixtures. It is important to note that the target area does not have to be physically entered to install a pickup device. Availability of a power supply is the major limitation of pickup devices. If the device can be installed so its electrical power is drawn from the available power within the target area, there will be minimal need for someone to service the device.

Transmission Links. Conductors carry the impulses created by the pickup device to the listening post. In lieu of conductors, the impulses can go to a transmitter that converts the electrical impulses into a modulated radio frequency (RF) signal for transmission to the listening post. The simplest transmission system is conventional wire. Existing conductors, such as used and unused telephone and electrical wire or ungrounded electrical conduits, may also be used. The development of miniature electronic components permits the creation of very small, easily concealed RF transmitters. Such transmitters may operate from standard power sources or may be battery operated. The devices themselves may be continuously operated or remotely activated.

Listening Posts. A listening post consists of an area containing the necessary equipment to receive the signals from the transmission link and process them for monitoring or recording. Listening posts use a receiver to detect the signal from a RF transmission link. The receiver converts the signal to an audio-video frequency and feeds it to the monitoring equipment. Receivers are small enough to be carried in pockets and may be battery operated. For wire transmission links only, a tape recorder is required. You can use many commercially available recorders in technical surveillance systems. Some of these have such features as a voice actuated start-stop and variable tape speeds (extended play). They may also have automatic volume control and be turned on or off from a remote location.

Telephone Monitoring

Monitoring telephone conversations is one of the most productive means of surreptitious collection of information. Because a telephone is used so frequently, people tend to forget that it poses a significant security threat. Telephones are susceptible to bugging and tapping.

A bug is a small hidden microphone or other device used to permit monitoring of a conversation. It may also allow listening to conversations in the vicinity of the telephone, even when the telephone is not in use.

A telephone tap is usually a direct connection to the telephone line permitting both sides of a telephone conversation to be monitored. Tapping can be done at any point along the line, for example, at connector blocks, junction boxes, or the multiwire cables leading to a telephone exchange or dial central office. Telephone lineman's test sets and miniature telephone monitoring devices are examples of taps. Indirect tapping of a line, requiring no physical connection to the line may also be accomplished.

The most thorough check is not absolute insurance against telephone monitoring. A dial central office or telephone exchange services telephone lines. The circuits contained within the dial central office allow for the undetected monitoring of telephone communications. Most telephone circuits servicing interstate communications depend on microwave links. Communications via microwave links are vulnerable to intercept and intelligence exploitation.

Miscellaneous

Current electronic technology produces technical surveillance devices that are extremely compact, highly sophisticated, and very effective. Miniaturized technical surveillance systems are available. They can be disguised, concealed, and used in a covert or clandestine manner. Variations of their use are limited only by the ingenuity of the technician. Equipment used in technical surveillance systems varies in size, physical appearance, and capacity. Many are identical to, and interchangeable with, components of commercially available telephones, calculators, and other electronic equipment.

Investigative Photography and Video Recording

Photography and video recording are offensive CI measures used to support intelligence and force protection. These are used in CI investigations for the following purposes:

- Identifying individuals. CI personnel perform both overt and surreptitious photography and video recording.
- Recording of incident scenes. CI personnel photograph overall views and specific shots of items at the incident scene.
- Recording activities of suspects. CI personnel use photography and video recording to provide a record of a suspect's activities observed during surveillance operations.

Polygraph

The polygraph examination is a highly structured technique conducted by specially trained CI personnel certified by proper authority as polygraph examiners. DOD Dir 5210.48, *DOD Polygraph Program*, provides guidance for the polygraph program generally. Within the Marine Corps there is no organic polygraph capability. If required, such support may be requested from the DIA via operational channels.

When Used

Do not conduct a polygraph examination as a substitute for securing evidence through skillful investigation and interrogation. The polygraph examination is an investigative aid and can be used to determine questions of fact, past or present. CI personnel cannot make a determination concerning an individual's intentions or motivations, since these are states of mind, not fact. However, consider the examination results along with other pertinent information available. Polygraph results will not be the sole basis of any final adjudication. The conduct of the polygraph examination is appropriate, with respect to CI investigations, only—

- When investigative leads and techniques have been completed as thoroughly as circumstances permit.
- When the subject of the investigation has been interviewed or thoroughly debriefed.
- When verification of the information by means of polygraph is deemed essential for completion or continuation of the investigation.

- To determine if a person is attempting deception concerning issues involved in an investigation.
- To obtain additional leads concerning the facts of an offense, the location of items, whereabouts of persons, or involvement of other, previously unknown individuals.
- To compare conflicting statements.
- To verify statements from witnesses or subjects.
- To provide a just and equitable resolution of a CI investigation when the subject of such an investigation requests an exculpatory polygraph in writing.

The polygraph examination consists of three basic phases: pretest, intest, and posttest.

Phases

During the pretest, appropriate rights advisement is given and a written consent to undergo polygraph examination is obtained from all examinees that are suspects or accused. If the examinee is a U.S. citizen or an alien lawfully authorized permanent residence status, advise him examinee of the Privacy Act of 1974 and the voluntary nature of examination. Conduct a detailed discussion of the issues for testing and complete the final formulation of questions to be used during testing.

During the intest phase, ask previously formulated and reviewed test questions and monitor and record the examinee's responses by the polygraph instrument. Relevant questions asked during any polygraph examination must deal only with factual situations and be as simple and direct as possible. Formulate these questions so examinee can answer only with a yes or no. Never use or ask unreviewed questions during the test.

If responses indicate deception, or unclear responses are noted during the test, conduct a posttest discussion with the examinee in an attempt to elicit information from the examinee to explain such responses.

Outcomes

A polygraph examiner may render one or more of four possible opinions concerning the polygraph examination:

- No Opinion (NO)—rendered when less than two charts are conducted concerning the relevant issues, or a medical reason halts the examination. Normally, three charts are conducted.
- Inconclusive (INCL)—rendered when there is insufficient information on making a determination.
- No Deception Indicated (NDI)—rendered when responses are consistent with an examinee being truthful regarding the relevant areas.
- Deception Indicated (DI)—when responses are consistent with an examinee being untruthful to the relevant test questions.

Factors Affecting Polygraph Results

Certain mental or physical conditions may influence a person's suitability for polygraph examination and affect responses during testing. CI personnel should report any information they possess concerning a person's mental or physical condition to the polygraph examiner before scheduling the examination.

Typical conditions of concern are—

- Mental disorders of any type.
- History of heart, respiratory, circulatory or nervous disorders.
- Current medical disorder, including colds, allergies or other conditions (such as pregnancy or recent surgery).
- Drug or alcohol use before the examination.
- Mental or physical fatigue.
- Pain or physical discomfort.

Conducting the Polygraph

To avoid such conditions as mental or physical fatigue, do not conduct prolonged or intensive interrogation or questioning immediately before a polygraph examination. CI personnel tell the potential examinee to continue taking any prescribed medication and bring it to the examination. Based on information provided by CI personnel and the examiner's own observations, the polygraph examiner decides whether or not a person is fit to undergo examination by polygraph. When CI personnel ask a person who is a U.S. citizen or alien lawfully authorized permanent residence status to undergo a polygraph examination, the person is told that the examination is voluntary and no adverse action can be taken based solely on the refusal to undergo examination by polygraph. Further, the person is informed that no information concerning a refusal to take a polygraph examination is recorded in any personnel file or record.

CI personnel will make no attempt to explain anything concerning the polygraph instrument or the conduct of the examination. If asked, they should inform the person that the polygraph examiner will provide a full explanation of the instrument and procedures before actual testing and that test questions will be fully reviewed with the potential examinee before testing.

Conduct polygraph examinations in a quiet, private location. The room used for the examination must contain, as a minimum, a desk or table, a chair for the examiner, and a comfortable chair with wide arms for the examinee. The room may contain minimal, simple decorations; must have at least one blank wall; and must be located in a quiet, noise-free area. Ideally, the room should be soundproof. Visual or audio monitoring devices may be used during the examination. However, if the examinee is a U.S. citizen or alien lawfully authorized permanent residence status, the examiner must inform the examinee that such equipment is being used and whether the examination will be monitored or recorded in any manner.

Normally, only the examiner and the examinee are in the room during an examination. When the examinee is an accused or suspect female and the examiner is a male, a female witness must be present to monitor the examination. The monitor may be in the examination room or observe through audio or visual equipment, if available.

On occasion, CI personnel must arrange for an interpreter to work with the examiner. The interpreter must be fluent in English and the required language, and have a security clearance appropriate to the classification of material or information to be discussed during the examination. The interpreter should be available in sufficient time before the examination to be briefed on the polygraph procedures and to establish the proper working relationship.

Miscellaneous

CI personnel will not prepare any reports concerning the results of a polygraph examination. This does not include information derived as a result of pretest or posttest admissions, nor include those situations where CI personnel must be called upon by the examiner to question the subject concerning those areas addressed before the completion of the examination.

7009. CI SURVEYS/VULNERABILITY ASSESSMENTS, EVALUATIONS, AND INSPECTIONS

Tactical Operations

During operations, CI surveys/vulnerability assessments, evaluations, and inspections, including TSCM inspections and surveys, are usually limited to permanent installations in rear areas or to key MAGTF C2 facilities. Purely physical security surveys and inspections do not fall under the cognizance of intelligence or CI personnel. Instead, these are conducted by trained physical security specialists under the purview of the MAGTF PMO. In those instances where perimeter security is the responsibility of a tactical unit, the physical security portion of the CI survey is primarily concerned with those areas within the perimeter containing classified material and areas susceptible to sabotage and terrorist attack. Special weapons sites require extra emphasis and may include CI monitoring of shipments in addition to other security services. Close and continuous liaison and coordination should be conducted with the local PMO during any CI survey, with the exception of a TSCM survey, to ensure complete coverage of the physical security aspects and any other areas with which the PMO may assist.

See appendix D for a CI survey/vulnerability assessment checklist and the format for a CI survey/vulnerability assessment report.

Garrison CI Inspections

Purpose

CI inspections are performed by commanders to determine compliance with established security policies and procedures. Commanders are responsible for security within their commands. These responsibilities, however, are executed by the command security manager with assistance from the unit intelligence officer, operations officer, communications and information systems officer, classified material control center custodian, and COMSEC material system custodian.

Access. CI personnel, while in the performance of their official duties are authorized access to all spaces (see MCO 3850.1H, *Policy and Guidance for Counterintelligence Activities*, for additional amplification).

Scope of the CI Inspection. The scope of the inspection will vary depending on its type and purpose. Inspections may include the following:

- Determine if assigned personnel with access to classified material are properly cleared.
- Determine if classified material is properly safeguarded by assigned personnel.
- Examine facilities and containers used for storing classified material to determine adequacy.
- Examine procedures for controlling entrances and exits, guard systems, and special guard instructions relating to security of classified material and sensitive areas.
- Examine the security and control of unit communications and information resources.
- Provide back brief to command security/intelligence personnel and formal results as required.

CI Credentials. CI credentials are intended for use of personnel on official CI missions. The CI Branch of the HQMC Intelligence Department is designated as the office of record for CI credentials. Credentials are not transferable and may be neither reproduced nor altered. Credentials are only issued to personnel who have completed a formal course of instruction in CI that qualified them for MOS 0204/0210/0211. Presentation of CI credentials certifies the bearer as having a TOP SECRET security clearance and sensitive compartmented information (SCI) access.

Personnel are directed to render assistance to properly identified CI personnel in the performance of their duties.

Types of Command Inspections

Announced Inspections. An announced inspection is one that has been publicized. Personnel concerned are aware of the inspection schedule and make preparations as necessary. Inspections are conducted on a recurring basis to ensure security practices meet established standards. The announced inspection is often accomplished with inspections conducted by the inspection staff of the common or a senior headquarters.

Unannounced Inspections. The unannounced inspection is conducted to determine compliance with security policies and procedures at a time when special preparations have not been made. The unit or section to be inspected is not informed in advance of the inspection. The inspection may be conducted at any time during or after normal working hours.

7010. CI SUPPORT TO THE CRISIS ACTION TEAM INTELLIGENCE CELL

When an intelligence cell is established in response to a terrorist threat or incident, CI personnel jointly man it with CID, NCIS, and if required, civilian law enforcement agents. The intelligence cell coordinates the intelligence, investigative, and criminal information needs of the installation and the on-scene operational commander. It should be separate from both the operations center and the crisis management force/on-scene commander but linked to both by a variety of wires and wireless communications means, including a direct data link. The design of the intelligence cell should be flexible to allow for the rapid integration of other federal, state, and local agencies, as appropriate. An intelligence cell may be established both in a garrison and a field environment.

These profiles are pertinent to CI support of any MAGTF unit executing these missions.

7011. CI MISSION PROFILES

The following CI mission profiles were initially developed to aid CI planning and execution in support of MEU (SOC) special operations missions.

Amphibious Raid

An amphibious raid is a landing from the sea on a hostile shore that involves swift incursion into or temporary occupancy of an objective and mission execution, followed by a planned withdrawal. Key CI requirements include—

- Assist the unit intelligence, operations and CIS officers with intelligence, CI, security and force protection planning.
- Provide threat intelligence to assist with planning and conducting MAGTF OPSEC activities.
- Assist with assessing security vulnerabilities and developing requirements; provide countermeasures recommendations.
- Establish access to CI and HUMINT data bases, automated links to JTF, other joint and services, coalitions, and host nation sources to help identify, assess, and develop countermeasures for threats.
- Conduct CI/HUMINT collection operations to satisfy tasked PIRs and IRs.
- Review CI data base for information related to enemy or other potential hostile PO&I, and then develop the CI target reduction plan.
- Develop/update CI threat estimates; assist intelligence officer with development/update of all-source intelligence estimate.
- Conduct liaison with U.S. embassy country team for third party assistance/escape and evasion.
- Attach CI personnel to raid force, when required, for document and material exploitation or on-scene debriefing/interrogation of friendly and/or enemy personnel in the objective area.
- Provide countersigns, challenges, and passwords.
- Conduct CI debrief of raid force; update CI files and data bases.

Limited Objective Attacks

- Assist the unit intelligence, operations and CIS officers with intelligence, CI, security and force protection planning.
- Provide OPSEC guidance.
- Develop/update CI threat estimates.
- Assess mission-oriented security vulnerabilities and develop requirements; provide countermeasures recommendations.

- Assist in the planning and conduct of counterreconnaissance operations to support the attack.
- Establish access to CI and HUMINT data bases, automated links to JTF, other joint and services, coalitions, and host nation sources to help identify, assess, and develop countermeasures for threats. Explore CI data base for information related to enemy or other potential hostile PO&I, and then develop the CI target reduction plan.
- Conduct CI/HUMINT collection operations (e.g., photographic reconnaissance) to satisfy tasked PIRs and IRs.
- Identify CI targets for possible exploitation and/or neutralization.
- Assist GCE and ACE intelligence officers with escape and evasion plans.
- Attach CI personnel to GCE when required.
- Conduct liaison with U.S. embassy country team for third party escape and evasion assistance.
- Provide countersigns, challenges, and passwords.
- Conduct CI debrief of assault force; update CI files and data bases.

Noncombatant Evacuation Operation

A noncombatant evacuation operation is conducted for evacuating civilian noncombatants from locations in a foreign country faced with the threat of hostile or potentially hostile action.

It will normally be conducted to evacuate U.S. citizens whose lives are in danger, but may also include the evacuation of U.S. military personnel, citizens of the host country and third country nationals friendly to the U.S. Key CI requirements include the following:

- Assist the unit intelligence, operations and CIS officers with intelligence, CI, security and force protection planning.
- Provide threat intelligence to assist with planning and conducting MAGTF OPSEC activities.
- Develop/update CI threat estimates.
- Assist with assessing security vulnerabilities and develop requirements; provide countermeasures recommendations.
- Assist in the planning and conduct of counterreconnaissance operations to support key sites.
- Establish access to CI and HUMINT data bases, automated links to JTF, other joint and services, coalitions, and host nation sources to help identify, assess, and develop countermeasures for threats. Explore CI data base for information related to enemy or other potential hostile PO&I, and then develop the CI target reduction plan.
- Provide recommendations and planning assistance regarding antiterrorism measures.
- Provide countersigns challenges and passwords.
- Provide CI officer and possibly a HST to the forward command element for on-scene liaison and support.
- Attach CI personnel to the CSSE evacuation control center to assist in time sensitive debriefs, liaison, antiterrorism measures, and to assist in personnel screening when directed.
- Provide from CI data base, a sanitized copy of the black, white, and gray lists to the CSSE evacuation control center for screening of persons of immediate interest.

- Conduct liaison with U.S. embassy country team for third party assistance.
- Conduct in-depth debriefs of non-combatant evacuees who may have information of intelligence/CI value.

Show of Force Operations

A show of force operation is designed to demonstrate U.S. resolve, which involves increased visibility of deployed military forces in an attempt to defuse a specific situation that may be detrimental to U.S. interests or national objectives.

- Assist the unit intelligence, operations, and CIS officers with intelligence, CI, security, and force protection planning.
- Assist and coordinate with unit psychological operations, public affairs, and civil affairs planners, with emphasis on development of operational plans to target and influence attitudes and behaviors of personnel within the AO.
- Develop/update CI threat estimates.
- Provide threat intelligence to assist with planning and conducting MAGTF OPSEC activities.
- Conduct liaison with U.S. embassy country team for third party assistance.

Reinforcement Operations

- Assist the unit intelligence, operations, and CIS officers with intelligence, CI, security, and force protection planning.
- Provide threat intelligence to assist with planning and conducting MAGTF OPSEC activities.
- Assist with assessing security vulnerabilities and develop requirements; provide countermeasures recommendations.
- Assist in the planning and conduct of counterreconnaissance operations to support key sites.
- Establish access to CI and HUMINT data bases, automated links to JTF, other joint and services, coalitions, and host nation sources to help identify, assess, and develop countermeasures for threats. Explore CI data base for information related to enemy or other potential hostile PO&I, and then develop the CI target reduction plan.
- Provide countersigns, challenges, and passwords.
- Attach CI personnel to GCE when directed to provide direct CI support.

Security Operations

- Assist the unit intelligence, operations, and CIS officers with intelligence, CI, security, and force protection planning.
- Provide threat intelligence to assist with planning and conducting MAGTF OPSEC activities.
- Provide recommendations and planning assistance regarding antiterrorism measures and countermeasures development.
- Provide estimates and recommendations on counterespionage and countersabotage countermeasures.

- Conduct CI/HUMINT collection operations to satisfy tasked PIRs, IRs, and EEFIs.
- Attach CI personnel to the GCE when directed for direct support to the GCE commander and to conduct special activities ashore.
- Assist in the planning and conduct of counterreconnaissance operations in the rear area.
- Establish access to CI and HUMINT data bases, automated links to JTF, other joint and services, coalitions, and host nation sources to help identify, assess, and develop countermeasures for threats. Research CI data base for information related to PO&I and development of the CI target reduction plan.
- Provide countersigns, challenges, and passwords.

Civic Action

Military civic action is the use of preponderantly indigenous military forces on projects useful to the local population in such fields as education, training, public works, agriculture, transportation, communications, health, sanitation, and others contributing to economic and social development, which would also serve to improve the standing of the military forces with the population.

- Provide threat intelligence to assist with planning and conducting MAGTF OPSEC activities.
- Establish access to CI and HUMINT data bases, automated links to JTF, other joint and services, coalitions, and host nation sources to help identify, assess, and develop countermeasures for threats.
- Provide terrorist and hostile intelligence service threat data.
- Provide countersigns, challenges, and passwords.
- Conduct debriefs in support of the foreign military intelligence collection activity (FORMICA) program.

Tactical Recovery of Aircraft and Personnel

Tactical recovery of aircraft and personnel (TRAP) is a mission performed by an assigned and briefed aircrew for the specific purpose of the recovery of friendly personnel, equipment, and/or aircraft when the tactical situation precludes search and rescue assets from responding and when survivors and their location have been confirmed. The mission is to expeditiously recover friendly aircrews or personnel in a wide range of political environments and threat levels. Key CI requirements include—

Equipment will either be recovered or destroyed, dependent upon the severity of the threat and environment, and the condition of the equipment.

- Assist the unit intelligence, operations, and CIS officers with intelligence, CI, security, and force protection planning.
- Provide countersigns, challenges, and passwords.
- Ensure isolated personnel report (ISOPREP) cards are up-to-date and readily accessible for all appropriate personnel prior to any operation. ISOPREP cards should be prepared and retained by either the unit security manager or its administrative officer. (See appendix J to JP 3-50.2, *Doctrine for Joint Combat Search and Rescue*, for the format and instructions for completing ISOPREP cards.)
- Conduct friendly POW/MIA investigations.

- Assist GCE and ACE intelligence officers and TRAP commanders in developing escape and evasion plans.
- Conduct liaison with U.S. embassy country team for third party escape and evasion and other support.
- Conduct CI debrief of TRAP force.

In-Extremis Hostage Rescue

- Assist the unit intelligence and operations officers with intelligence, CI, and force protection planning.
- Provide countersigns, challenges, and passwords.
- Attach CI personnel to the in-extremis hostage rescue (IHR) strike element when directed for target exploitation and personnel handling.
- Provide threat intelligence to assist with planning and conducting MAGTF OPSEC activities.
- Assist in rapid planning.
- Arrange for in-extremis hostage rescue IHR force isolation.
- Conduct on-scene document and material exploitation when directed.
- Conduct initial hostage/terrorist debriefs.
- Provide assistance in training the IHR force in urban surveillance and counter-surveillance.
- Conduct liaison with national and theater intelligence agencies on hostage rescue and HUMINT operations.

CHAPTER 8. COUNTERINTELLIGENCE TRAINING

8001. GENERAL

The effectiveness of CI and force protection security measures often rests on the individual Marine's ability to recognize and accurately report threats to the security of the command. It also rests on the Marine's willing acceptance of a high degree of security discipline.

Training Objective

The ultimate objective of CI training ensures effective contribution by MAGTF personnel to the CI effort and instills a sense of security discipline. To ensure that the individual Marine can support effective CI and security measures, CI training is integrated with other intelligence and command training programs. For CI personnel, the objective of CI/HUMINT training ensures they are capable of providing the CI/HUMINT support required by the commander. Additionally, it ensures non-MAGTF CI organizations and capabilities are understood and prepared to effectively integrate with and support MAGTF operations.

Generally, training can be divided into the following categories:

- Basic CI and security training for all personnel.
- Training for CI personnel.
- CI training for intelligence personnel.
- Mission-oriented CI training.

Basic CI Training

Basic CI and security training requirements are common to commands. However, emphasis on certain subjects will vary according to the mission of the command and the duty assignments of personnel within the unit.

8002. BASIC CI AND SECURITY TRAINING FOR ALL PERSONNEL

All personnel receive training in CI and force protection to safeguard friendly forces from the hostile intelligence threat. The following related areas should be covered to instill awareness on the part of all personnel:

- Operations security, including its purpose, how to identify unit/individual patterns and profiles that can be exploited by the enemy, and countermeasures to minimize or eliminate these.
- Information security, including levels of security classification, when to apply a security classification, ramifications of security classifications on friendly operations, and development of operational and functional classification guidance and criteria for downgrading and declassification.
- Personnel security, including individual standards, how to identify risks and vulnerabilities, and the individual and command actions to take when risks and vulnerabilities are identified.
- Purpose and procedures for the use of countersigns, challenges, and passwords.
- Survival, evasion, resistance escape (SERE), including training on prospective AO, the nature and attitude of the civilian populaces, and the techniques and procedures used by threat forces.

DOD Dir 1300.7, *Training and Education Measures Necessary to Support the Code of Conduct*, provides policy and guidance on SERE training. It establishes three levels:

Level A—the minimum level of understanding required of all Armed Forces personnel, to be provided during entry level training.

Level B—the minimum level of understanding required of military personnel whose military occupational specialities and assignments entail moderate risk of capture. Level B training is to be conducted as soon as possible upon assumption of the duty/ MOS that makes them eligible.

Level C—the minimum level of understanding required of military personnel whose MOS/assignment entails significant risk of capture, or whose position, rank, or seniority make them vulnerable to greater than average targeting/exploitation by enemies or other threats. Examples include aircrews, ground reconnaissance personnel, and military attaches.

- Communications and information security, including vulnerabilities and countermeasures.
- U.S. code of conduct.
- Identity of the unit security manager and other personnel with leadership roles regarding unit security including—
  - Unit security manager—overall coordination of unit security. (The unit security manager is generally either the unit's chief of staff or executive officer.)
  - G-1/S-1—principal staff cognizance for classified materials control.
  - G-2/S-2—principal staff cognizance for sensitive compartmented information and special security, and identification of enemy intelligence capabilities and operations.
  - G-3/S-3—principal staff cognizance for force protection, C2, operations security, counter-reconnaissance, deception, and electronic protection.
  - G-6/S-6—principal staff cognizance for CIS security, cryptographic materials system.
  - Headquarters commandant—principal staff cognizance for physical security.
- Briefs regarding attempts or acts of espionage, subversion, terrorism or sabotage should be emphasized. Individual responsibilities for reporting foreign contact, perceived or actual attempts at espionage or subversion, and undue interest on the part of anyone to acquire terrorist countermeasures or intelligence collection information should be stressed regardless of MOS. Routine threat awareness should become a part of each person's professional military education objectives. This is most easily satisfied at formal professional military education courses or correspondence courses but can also be obtained during mission-oriented training. (See DODINST 5240.6, *CI Awareness and Briefing Program*, for additional information.)

8003. TRAINING FOR OFFICERS AND SNCOS

Officers and SNCOs must receive training in the following CI and related security subjects.

- Identification, control, and reporting of persons and installations of CI interest.
- Methods of operations and capabilities of hostile organizations within the AO.
- Operations security as a process of analyzing friendly actions attendant to military operations and other activities to—
  - Understand, identify, and properly employ the use of EEFI.
  - Identify those friendly actions and operational patterns that can be observed and exploited by enemy intelligence forces.
  - Determine indicators that hostile intelligence elements might obtain. These indicators could be interpreted or pieced together to derive critical information in time to be useful to adversaries.
  - Select, plan, and execute friendly protective measures that eliminate or reduce to an acceptable level the vulnerabilities of friendly actions to adversary exploitation.

- Protection of classified material and other information that may be of value to the enemy. Personnel must be able to name and define the three levels of security classification, minimum-security standards, and the potential damage that may be caused if this information should be exposed to unauthorized persons. Officers and SNCOs must in particular understand how to apply these to their activities and products, as within the specific exercise/operation classification guidance.
- Evaluation of the suitability of subordinate personnel who have access to classified information. Officers and staff noncommissioned officers must be able to recognize indicators associated with potential involvement or susceptibility to espionage activities, such as unexplained affluence, erratic behavior, or mood swings, and then initiate action per OPNAVINST 5510.1.
- Methods of operation and acquistion of information, hostile intelligence services organizations, and capabilities. Additionally, briefs on current events concerning attempts and/or acts of espionage, subversion, terrorism, or sabotage should be emphasized. Individual responsibilities for reporting foreign contact, perceived or actual attempts at espionage or subversion, and undue interest on the part of anyone to acquire terrorist countermeasures as well as intelligence collection operations should be stressed regardless of MOS. Routine threat awareness should become a part of each person's professional military education objectives. This is most easily satisfied at formal professional military education courses or correspondence courses, but can also be obtained during mission-oriented training. Special attention should be paid to indications of the level of terrorist/subversive activity through unclassified or classified articles/ publications and through tailored briefs, particularly in those countries identified as high threat areas. (See DODINST 5240.6 for additional information.)
- Use of countersigns, challenges, and passwords.
- Purpose, scope, organization, capabilities, and limitations of Marine Corps CI assets.
- Handling of personnel of CI interest during MAGTF operations, including the identification, control, reporting of persons, installations and materials of CI interest.

8004. MISSION-ORIENTED CI TRAINING

General

Mission-oriented training ensures that the unit's objectives are achieved by employing the proper CI measures. Unit SOPs and training exercises should include the following:

- Operations security measures and passive countermeasures to protect sensitive or classified information from the enemy or unauthorized personnel.
- Counterreconnaissance activity to prevent observation from opposing forces, such as patrols, camouflage, and other measures.
- Other security measures designed specifically for each type of unit and the nature of the operation.

- CIS security procedures designed to lessen friendly signatures and susceptibility to hostile radio electronic combat operations.
- SERE training for those personnel whose military jobs, specialties, or assignments entails moderate risk of capture.

A complete listing and description of the intelligence training standards for CI personnel can be found in MCO 3500, *Training and Readiness Manuals* series.

CI Personnel

The following identifies the principal individual training standards.

- Supervise CI/HUMINT Co and staff CI/HUMINT section garrison activities.
- Supervise CI/HUMINT Co and staff CI/HUMINT section in a tactical environment.
- Supervise CI teams/HUMINT support teams headquarters activities and operations.
- Monitor CI training plan for CI personnel.
- Prepare the CI SOP.
- Brief CI/HUMINT missions, tasks, and authorizing directives and regulations.
- Conduct CI screening.
- Conduct mobile and static checkpoints.
- Conduct CI activities in support of cordon and search.
- Conduct CI survey.
- Conduct missing in action investigation.
- Conduct an investigation of an act of espionage, sabotage, subversion or terrorism.
- Conduct CI surveillance.
- Conduct CI countersurveillance.
- Conduct CI interrogation.
- Conduct map tracking during interrogation and interviews.
- Exploit captured documents and equipment.
- Conduct CI elicitation.
- Conduct CI debrief.
- Conduct CI interview.
- Conduct CI/HUMINT liaison.
- Account for operational funds.
- Conduct technical surveillance countermeasures service.
- Maintain CI and HUMINT equipment.
- Operate current automated intelligence systems.
- Provide CI support to MAGTF operations.
- Conduct CI activities in support of noncombatant evacuation operations.

8005. TRAINING OF INTELLIGENCE SECTION PERSONNEL

The following subjects are considered appropriate for the CI training of intelligence section personnel and should incorporate MAGTF, other services, joint and national capabilities, issues and operations:

- CI collection, processing, analysis and production, and dissemination organizations, capabilities and limitations.
- C2 and CIS architecture. C2 and supporting CIS operations, both for internal CI activities and for overall integrated CI/intelligence operations.
- CI sources of information and methods of reporting. Walk-ins, host nation liaison activity, line crossers and CFSO are examples of sources utilized in CI/HUMINT operations to support command intelligence objectives.
- CI support activities, to include CI surveys/vulnerability assessments, technical support and TSCM.
- Intelligence oversight. When intelligence specialist assets are attached or placed in direct support, the minimum reporting requirements and prohibited activities should be strictly monitored and enforced per DOD 5240.1, *DOD Intelligence Activities*.
- Unique supplies, embarkation, maintenance, and other functional support to CI.

8006. PEACETIME CI TRAINING

Exercises

The use of exercise CI provides commanders, staffs, and units involved realistic experience planning and executing CI operations, in working with CI information, rules, communications, and personnel, and in using CI when planning and executing operations. Exercise CI operations may be conducted whether or not an opposition force exists.

Exercise CI may be scripted or preplanned if an opposition force does not participate in the exercise, such as for a staff exercise or a command post exercise (CPX). Use of scripted CI should be planned well in advance of the exercise to allow adequate time to script the exercise CI necessary to realistically support the exercise scenario. Coordination between exercise planners and exercise CI scripters is important to ensure that the exercise CI reporting simulates realistic CI activities, information, intelligence dissemination flow and timelines (e.g., the time-sensitive limitations of CI for timely reporting and access to sources often do not integrate well with accelerated wargaming time clocks) in relation to the notional opposition force. Security requirements, such as for the use of special CI communications, must be maintained throughout the exercise.

Conduct of CI operations during exercises is closely controlled. CI operations require the same security precautions and controls for exercises required for real-world operations.

Exercise CI operations may also be conducted against a live opposition force, such as during a MAGTF field exercise. This provides for more realistic training for both the CI element and users of SIGINT participating in the exercise, as well as better CI/other intelligence element integration and training. Depending on the level of the exercise, use of simulators and national systems may be requested for participation in the exercise to add realism and enhance training.

Real-World Support

During the conduct of exercises, particularly overseas, CI personnel provide critical, real-world support to the unit's force protection mission. This support involves protecting the force prior to and during training from exposure to or exploitation by hostile intelligence and security services, and terrorist actions targeting the force.

8007. CI TRAINING PROGRAMS

Personnel receive training in CI and security as a basis for fulfilling their basic responsibilities to safeguard information of value to the hostile intelligence threat. CI personnel receive additional training to improve their proficiency in accomplishing the CI mission.

Individual CI Personnel Training

The training of CI personnel is driven by MCO 3500 series training and readiness manuals. Training standards are derived from mission performance standards. Mission performance standards are further derived from the combat requirements of the operating forces and establish a common base of training for Marines who have the same MOS.

Responsibilities

The following personnel and organizations have the responsibility for ensuring that a viable program is established for CI training:

- CI resident course—Navy and Marine Corps Intelligence Training Center (NMITC).
- Advance training community—Intel bn and CI/HUMINT company commanders, and HQMC Intelligence Department.

Advanced training includes, but is not limited to, the following:

- Intelligence cross training.
- TSCM training.
- Photographic, video and electronic surveillance systems training.
- Military Officer Training Course/Military Officer Familiarization Course/ Officer Support Specialty Course (MOTC/MOFC/OSSC).
- SERE training.
- Terrorism/ counterterrorism training.
- Intelligence and CI communications and information systems training.

Descriptions

CI Resident Formal School Training

CI resident entry-level formal school training for both officers and enlisted Marines is via the 17-weeks MAGTF CI Agents Course conducted at the NMITC, Dam Neck, VA. Successful completion of this course provides basic MOS qualification (officers—MOS 0204/enlisted—MOS 0211) and certification as Level I Anti-Terrorism/Force Protection Instructors. Once qualified, CI personnel are required to maintain proficiency in those training standards achieved.

Advanced Training

Advanced training of CI personnel in specialized skills is conducted to enhance their abilities to perform increasingly complex tasks. This training supplements and is conducted within the post resident training process.

CHAPTER 9. COUNTERINTELLIGENCE ADMINISTRATION

9001. GENERAL

Administration of CI elements consists of files, reports, communications, and emergency funds. CI establishes and maintains operational files essential to their combat CI mission. The accomplishment of the CI mission requires accurate, timely, and pertinent reports disseminated in a usable form. CI has organic communications equipment to help coordinate CI activities and report information to other organizations. Emergency and extraordinary expense (E&EE) funds are made available for CI because of the nature of the missions.

9002. FILES

The following operational files are normally maintained in a combat environment by CI elements at all echelons. Formats, organization, and content for each should be coordinated with the P&A cell OIC or the supported unit's intelligence officer.

- Information concerning personalities, organizations, installations, and incidents of current and future CI interests. Often basic information of this type is recorded in a card file/folder or automated data base for ready reference. It is also cross-indexed to more detailed information.
- Correspondence and reports about specific operations and investigations.
- Source records containing essential data on sources of information.
- Area files containing basic reference data and information on enemy intelligence activity and CI measures within a particular geographic area.

9003. REPORTS

The method of dissemination of CI information depends primarily on the nature and urgency of the information, the location of the receiving units, the security requirements, and the means available.

CI reports transmit accurate information to units to support planning, decisionmaking, and execution; aid in the processing of intelligence; and serve as a record of CI activities. Normally, information is disseminated by record or voice messages, personal liaison, telephone, briefings, messenger, and written reports. CI reports will be written per the formats prescribed for a standard naval letter or message. The report formats in appendix D are DOD standard formats meant to enhance joint interoperability. They should not be modified unless absolutely necessary and following coordination with pertinent intelligence organizations. Reports are classified according to content.

9004. PERSONNEL

Augmentation

When additional CI personnel are needed, the requirement is identified by the unit intelligence officer to the personnel officer for validation by the commander. The request will then be forwarded through the chain of command to the MEF G-2 for validation, prioritization, and follow-on tasking to the intel bn.

Global Sourcing

When the MEF's organic CI resources are insufficient to fulfill a validated requirement, it is then forwarded to HQMC for global sourcing support from either the other MEFs or Marine Corps Forces Reserve (MARFORRES).

Reserves

There are three reserve CI teams within MARFORRES. The 10th and 12th CI teams are located at Anacostia Naval Air Station, Washington, D.C. The 14th CI team is located at Miramar Naval Air Station, San Diego, CA.

9005. EMERGENCY AND EXTRAORDINARY EXPENSE FUNDS

The nature of certain CI and intelligence activities is such that security considerations, opportunity, timeliness or other circumstances may make the use of normal military funds impractical or undesirable. The Secretary of the Navy has authorized the use of E&EE funds for certain intelligence and CI activities.

Subhead 123.a funds are General Defense Intelligence Program (GDIP) monies intended for use by Naval Attaches in the performance of their official duties and are managed by the Office of Naval Intelligence and coordinated by CMC.

Subhead 123.b funds are FCIP monies intended for CI functions only and are managed through NCIS.

Intelligence collection funds to support HUMINT operations conducted by Marine Corps CI assets are available through E&EE (subhead 123.a) funds. These funds are not authorized for the conduct of CI activities.

CI funds to support offensive and defensive CI operations are available through E&EE (subhead 123.B) funds. These funds are not authorized for the conduct of positive HUMINT or other controlled intelligence collection activities.

The MAGTF CIHO initiates early planning and coordination to ensure the availability of both types of funds. This function is performed by the CI detachment or HST OICs within MEU (SOC)s or SPMAGTFs.

See MCO 7040.10A, *Emergency and Extraordinary Expense Funds*, for additional information on the use, control, and accounting of E&EE funds.

CHAPTER 10. GARRISON COUNTERINTELLIGENCE SUPPORT

10001. MISSION

The primary garrison mission of CI activities is planning, preparing, and training to accomplish MAGTF CI functions and operations. A secondary mission is to advise and assist the commander in implementing the command's force protection and security programs and supporting command initiated security measures. CI is designed to identify and neutralize the effectiveness of both potential and active hostile collection efforts and to identify and neutralize the effectiveness of individuals, activities or organizations capable of engaging in hostile intelligence collection, sabotage, subversion or terrorism directed against the command.

Additional doctrine pertaining to countering terrorism is contained in MCO 3302.1, *USMC Antiterrorism Program.*

10002. COUNTERINTELLIGENCE SURVEY/VULNERABILITY ASSESSMENT

Basis

The CI survey/vulnerability assessment is designed to assist commanders in establishing security systems, procedures, and safeguards to protect military personnel, organizations, and installations from espionage, sabotage, terrorism or subversion. The survey assesses a unit's overall security posture against threats identified in the CI estimate. The CI survey/vulnerability assessment will identify specific vulnerabilities to hostile intelligence, espionage, sabotage, and subversion or terrorist capabilities and provide recommendations on how to eliminate or minimize these vulnerabilities. It is necessary that the survey/vulnerability assessment look forward in both space and time to support the development of CI measures necessary to protect the unit as it carries out successive phases of the operation. The CI survey/vulnerability assessment includes—

The survey/vulnerability assessment is not a recurring event. Once it is conducted, the survey/vulnerability assessment will remain valid for that specific installation or facility until there are major changes in the physical security, the mission of the command, or potential threats.

- Analysis of CI factors influencing security within the unit or installation.
- A determination of CI measures required by the sensitivity or criticality of the installation.
- An assessment of CI measures and deficiencies that currently exist and their effectiveness.
- Recommendations for improvements to these measures or the initiation of new security measures to achieve required security standards and protection.

Initiation

Initiation of a CI survey/vulnerability assessment begins with a request from the commander of a unit or installation concerned or with a higher commander in the same chain of command.

That request will normally occur under the following circumstances:

- Changes in known or estimated threat risks.
- Activation or reactivation of an installation or a major command.
- Significant change in the mission, functions or physical reorganization of an installation or major command.
- New hazardous conditions affecting an installation that necessitate the reevaluation of the security systems in place.
- Significant changes in the level/scope of classified material stored, handled, processed, and/or produced.
- Change in locale or environment that the installation is located.

Preparation

When preparing to conduct a CI survey/vulnerability assessment, there are four areas that need to be considered: selection of personnel, collection of data, coordination, and the preparation of checklists. The scope and depth of each of the areas to be considered will depend entirely on the unit or installation itself. The following paragraphs offer some ideas.

Selection of Personnel

Selection of personnel will consider the number of persons required to complete the task and available assets and should include TSCM personnel, if possible.

Collection of Data

At a minimum, data collected should include the mission, organization, functions, and security directives pertinent to the installations. Reports of previous CI surveys/vulnerability assessments, inspections or evaluations should be acquired and reviewed.

Coordination

Coordination with the commander of the unit or installation should be conducted by a CI officer/specialist. This coordination will help in determining the scope of the survey/vulnerability assessment, arrange for access to required records or areas, procurement of directives if not already acquired, arrangements for any required escorts, and arrangements for necessary briefings.

Preparation of Checklist

Checklist preparation will evolve when a thorough review of the commander's objectives and the unit/installation's mission, organization, and operations has been completed. The checklist includes general and specific points to be covered during the survey/vulnerability assessment. It also serves as a reminder to the surveying personnel to satisfy the predetermined scope of the survey/vulnerability assessment. In an effort to assist in formulating a checklist, areas of emphasis typically include document security, personnel security, communications and information systems security, and actual physical security requirements (less those physical security requirements falling under the purview of the provost marshal). Physical security requirements should be coordinated with the

PMO because the PMO has resident knowledge and expertise as the primary agency for physical security aboard the installation. From those three general points, more specific points will evolve.

A comprehensive CI survey/vulnerability assessment checklist is contained in appendix D.

Conduct

The actual conduct of the CI survey/vulnerability assessment will depend solely on the findings set forth in the data collected and the needs developed in the checklist(s). The judgment of the leader will determine just how/what the team will do. There are several things that should be noted as specific methods and/or ways to conduct the CI survey/vulnerability assessment including—

- The survey/vulnerability assessment team should make a physical tour of the installation from its perimeter areas to the center, including the area immediately outside its physical boundary using one possible concept of concentric circles. This tour should include every building, area, facility, or office requiring special security considerations or considered sensitive. It is also recommended that unit/installation staff personnel and subordinate commanders are interviewed, as required, to assist in determining the operational importance, and known vulnerabilities and security practices of each area surveyed.
- The cost of replacement, (in terms of time and not necessarily dollars and cents), of personnel, documents, and materials in the event the installation is neutralized or destroyed. The potential sources for the procurement of comparable personnel and sources for copies of essential and critical documents to replace or reactivate the installation.
- The location of the unit/installation and the effects of the surrounding environment/elements on the overall security of the installation.
- The level of classified and sensitive information used, produced, stored, or compiled.
- The criticality of the installation with the overall defense posture of the U.S. based on its mission/function.
- Whether there are other units/installations/facilities that can assume the role of the surveyed unit/installation if it is neutralized or destroyed.
- The unit/installation's vulnerability to terrorist or special operations forces attacks based on local/international threat and conditions.

Baseline

Once the level of security required has been determined, the posture and effectiveness of existing security measures must be assessed. Areas that should be examined include—

- **Document Security.** Document security is systematic review and inspection of all security procedures and records used in the handling of classified documents, information, and other classified material. The review should include the flow of classified material beginning from its creation/receipt at the installation/unit/command to its final storage area or destruction.
- **Personnel Security.** Personnel security is based on the relationship existing between a unit/installation's mission, local CI threat estimates,

the actual security level of assigned personnel, and the supporting security education and awareness program.

- **Physical Security.** This assesses the system of security controls, barriers, and other devices and procedures to prevent destruction, damage, and unauthorized access to the installation and facilities. In accordance with MCO 3302.1, this is normally the responsibility of the provost marshal. However, with the storage of classified material where the susceptibility to espionage, sabotage, subversion or terrorism is a consideration, the CI survey/vulnerability assessment is applicable. Whether the installation is a controlled one or an open post, the actual physical requirements established by directives will be examined based on the unit/installation's mission and nature/type materials being used. Emphasis is on the examination of the physical security factors affecting classified storage areas, security areas, critical areas that require protection from sabotage or terrorist attack, and other locations that may be designated as sensitive.

See appendix D for the format of a CI survey/vulnerability assessment report.

Exit Brief

Once the survey/vulnerability assessment has been completed and recommendations have been formulated, the survey/vulnerability assessment team will provide the unit/installation commander with an exit brief addressing preliminary findings and recommendations. Compliance will then be the commander's responsibility.

CI Survey/Vulnerability Assessment Report and Recommendations

Once the CI survey/vulnerability assessment has been completed and data compiled, a formal report of findings will be written and recommendations made. Recommendations will be based on the security measures required of the command, existing measures, and procedures. Recommendations provide measures to safeguard the installation/organization against sabotage, espionage, subversion, and/or terrorism. Each recommendation will be in response to a specific identifiable hazard with consideration given to cost, time, manpower, and availability of materials. If at all possible, alternate recommendations should be included.

10003. COUNTERINTELLIGENCE PENETRATION INSPECTION

Once a survey/vulnerability assessment has been conducted on a unit/installation/facility, a CI penetration inspection may be conducted to determine the effectiveness of recommendations implemented. The inspection is designed to provide a realistic test of established security measures and practices. It is conducted in a manner that installation personnel, other than the commander and those persons informed, are unaware that such an action is taking place. The inspection may be all-inclusive or may be limited to an attempt by CI personnel to fraudulently gain access to specific sensitive areas for performing simulated acts of espionage or sabotage. These simulated acts should be as realistic as possible. These acts should correspond to activities that could be attempted by area threats or hostile agents. The penetration inspection must be thoroughly planned and coordinated and include the following considerations:

- A responsible person, who is knowledgeable of the inspection and a representative of the inspected command, must be present during the inspection.
- In addition to the CI credentials and military identification, inspectors must carry a letter of identification and authorization for use only in emergency situations.
- Termination of the inspection will be done immediately if at any time personnel are subject to physical danger or other safety risks.
- Preparation for and conduct of the inspection must not impair or disrupt the normal operation/function of the command unless the inspection is specifically designed to do so.
- Command or installation personnel will not be used in any manner that would tend to discredit them.

10004. COUNTERINTELLIGENCE EVALUATION

CI evaluations are similar to surveys but are limited in scope. The CI evaluation is normally conducted for a small unit or a component of a larger organization when there has been a change in the security posture, an activation or reactivation of a facility, a physical relocation or substantive changes to the unit's facilities or CIS infrastructure. CI evaluations are normally limited to areas containing or processing classified material.

The CI evaluation may be limited to an assessment of one type of security, (e.g., document, personnel or physical security) or it may include any combination, depending on the needs of the unit. The procedures for the preparation and conduct of the evaluation are the same as those for the CI survey/vulnerability assessment. However, the procedures usually are not as extensive. The CI evaluation also may be used to update CI surveys when only minor changes have occurred within an installation or major organization.

10005. TECHNICAL SURVEILLANCE COUNTERMEASURES SUPPORT

TSCM operations are governed by DOD Directive 5200.9 and SECNAVINST 5500.31.

As discussed in chapter 7, the purpose of the TSCM program is to locate and neutralize technical surveillance devices that have been targeted against U.S. sensitive or secure areas. CI TSCM teams have specialized equipment and techniques to locate and identify threat technical surveillance activity. TSCM support consists of inspections and surveys. A TSCM inspection is an evaluation to determine the physical security measures required to protect an area against visual and audio surveillance. TSCM surveys include a complete electronic and physical search for unauthorized modification of equipment, the presence of clandestine audio and visual devices, and other conditions that may allow the unauthorized transmission of any conversation out of the area being surveyed.

Historically, hostile intelligence services have used technical surveillance monitoring systems in their intelligence and espionage operations against U.S. targets, both in the continental U.S. and abroad. A technical surveillance monitoring system may be defined as any visual surveillance or audio monitoring system used clandestinely to obtain classified or sensitive unclassified information for intelligence purposes. These monitoring systems include, but are not limited to, the following:

- Sound pickup devices, such as microphones and other transducers that use wire and amplifying equipment.
- Passive modulators.
- Energy beams, i.e., electromagnetic, laser, and infrared.
- Radio transmitters.
- Recording equipment.
- Telephones, i.e., taps and bugs.
- Photographic and television cameras.

See chapter 7, paragraph 7008, for additional information on TSCMs.

Requests for TSCM support must be classified and no conversation concerning the inspection should take place in the vicinity of the area to be inspected. Procedures for requesting inspections and surveys, TSCM responsibilities, and further information on the audio surveillance threat are contained in OPNAVINST 0500.46 and MCO 5511.11.

The CI platoon of each CI/HUMINT Co has one TSCM team to support the MEF's requirements. This capability is designed primarily for combat support but also supplements the NCIS TSCM responsibilities in garrison.

Appendix A. Counterintelligence Principal and Supporting Equipment

MARINE CORPS COMMON EQUIPMENT

The CI detachment or HST is the basic building block for CI HUMINT support to support a MAGTF or subordinate unit. The HST reports to the supported commander with their authorized organic equipment under the table of equipment (T/E) 4714 series. This generally includes at a minimum, but is not limited to, the following organic Marine Corps common equipment for each three-man element and would require two sets to fully equip an HST.

Qty	Description
1	M998, high mobility multipurpose wheeled vehicle (HMMWV) complete with SINCGARS radio mount
1	Trailer, cargo, 3/4 ton, two-wheel, M101A3
1	Command post (CP) tent, with applicable support poles
2	Radar scattering nets, with applicable support poles
1	Records chest
1	Lantern chest with
2	lanterns and stove
1	SINCGARS Radio, Radio Set, AN/PRC-140B
1	Radio Set, AN/PRC-119A
1	Navigation Set, Satellite (PLGR) AN/PSN-11
3	Sleeping cots
1	six-cube box containing, stools, extension cords, supplies etc.

CI/HUMINT Equipment Program

In addition to that equipment officially on CI/HUMINT Co's T/E, the CI/HUMINT Company maintains a special allowance account of CI unique equipment. This CI/ HUMINT equipment program (CIHEP) allotment provides increased capabilities for conducting CI operations in an urban or non-tactical environment.

The CIHEP allowance is continuously upgraded. The following items are currently included.

Qty	Description
3	Motorola SABER, receiver-transmitter
3	Antenna, magnetic mount
1	Motorola 20 watt base station/repeater
1	Digital encryption loader (DES)
2	Motorola SABER recharger bank
6	SABER batteries
1	Kodak DCS-420 digital camera set

Qty	Description
1	CI/HUMINT automated tool set (CHATS) containing a notebook computer, color printer, color scanner, DC-50 digital camera and secure communications and FAX capability.
1	AT&T 1100 STU-III telephone
1	Tripod
1	Camera, Hi-8mm video
1	Video capture card
1	Video, Hi-8mm, TV recorder/player, five-inch screen
1	Video, Hi-8mm, TV recorder/player, two-inch screen
2	Recorder, microcassette
1	Metal detector

CHATS CURRENT CAPABILITIES

CHATS is a suite of hardware designed to meet the unique requirements of MAGTF CI/HUMINT elements (see fig. A-1). Authorized to operate up to the SECRET level and using the baseline and DCIIS software suite, the system provides the capability to manage assets and analyzes information collected through investigations, interrogations, collection, and document exploitation. With CHATS, MAGTF CI elements may electronically store collected information in a local data base, associate information with digital photography, and transmit/receive information over existing military and civilian communications.

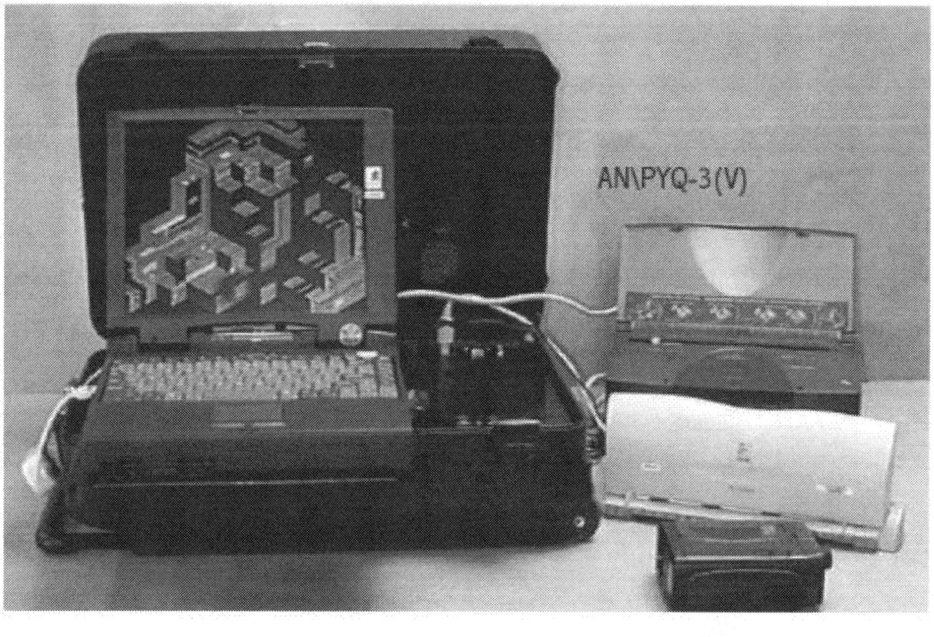

Figure A-1. CI/HUMINT Automated Tools Set.

CHATS provides these functions primarily with commercial off-the-shelf software operating in a laptop computer within a hardened transport case. Major systems components include—

Operating System:	MS Windows 95/MS Plus for Windows 95
Hardened System:	Intel Pentium 166 MHz or faster
Disk Drive:	1.3 GB removable hard drive
CD-ROM:	12X
RAM:	32 MB
Communications:	STU-III (AT&T 1100) (unit provided) or secure terminal equipment
Secure FAX:	Ilex PCMCIA
Digital Camera:	Kodak color DC50
Printer:	Cannon BJC-70
Color Scanner:	Logitec PowerPage
External Modem:	PCMCIA 33.6 BPS
Comms Paths:	Ethernet Thin LAN, Commercial Telephone, and CNR

When fully fielded, the system will enhance seamless integration of CI/HUMINT information from the HST to intelligence units and sections throughout a MAGTF. Current planning envisions the capability to be Global Command and Control System compliant and able to support the information exchange between CHATS and the Marine Corps IAS, the army's all source analysis system (ASAS), and the joint community's JDISS.

APPENDIX B. COUNTERINTELLIGENCE APPENDIX (APPENDIX 3 TO ANNEX B, INTELLIGENCE)

CLASSIFICATION

Copy no. ___ of ___ copies
OFFICIAL DESIGNATION OF COMMAND
PLACE OF ISSUE
Date/Time Group
Message reference number

APPENDIX 3 TO ANNEX B TO OPORD XXX (U)

COUNTERINTELLIGENCE (U)

(U) REFERENCES: Identify DOD, DIA, CIA, and other directives; combatant commander, JTF, or other higher authorities' operations plans, orders and tactics, techniques and procedures or SOP for intelligence and CI operations; pertinent maps and other geospatial information resources; and any other relevant references that pertain to anticipated MAGTF CI operations.

1. (U) General

a. (U) Objectives. Discuss general objectives and guidance necessary to accomplish the mission.

b. (U) Command Responsibilities and Reporting Procedures. Provide a general statement of command responsibilities and reporting procedures to ensure the flow of pertinent CI information to higher, adjacent, or subordinate commands.

c. (U) CI Liaison Responsibilities. Discuss responsibility to coordinate and conduct liaison between command CI elements and those of other U.S. and allied commands and agencies.

d. (U) Restrictions. Discuss the effect of U.S. Statutes, Executive Orders, DOD and Higher Headquarters Directives, and SOFA on CI activities.

2. (U) Hostile Threat. Refer to Annex B and current intelligence estimates for threat capabilities, limitations, vulnerabilities, and OOB pertinent to CI operations. Summarize the foreign intelligence activity and collection threat; foreign security and CI threat; and threats from sabotage, terrorism, and assassination directed by foreign elements. Emphasize capabilities, limitations, and intentions. Ensure that at a minimum the most likely and worst cases are addressed.

3. (U) Mission. State concisely the CI mission as it relates to the MAGTF's planned operation.

Page number

CLASSIFICATION

4. (U) Execution

a. (U) Concept of Operations. Reference the unit's intelligence SOP and Appendix 16 (Intelligence Operations Plan) to Annex B. Restate as appropriate the commander's intent and pertinent aspects of the unit's overall concept of operations as they relate to CI operations. Outline the purpose and concept of CI operations, specified priorities, and summarize the means and agencies to be employed in planning and directing, collecting, processing and exploiting, analyzing and producing, disseminating, and using CI during execution of the OPORD. Address the integration of JTF, other components, theater, national and allied forces' CI operations.

b. (U) Tasks for CI and Related Units and Organizations, Subordinate Units, and Task Force Commanders/OICs.

(1) (U) Orders to Subordinate, Attached, and Supporting Units. Use separate numbered subparagraphs to list detailed instructions for each unit conducting CI operations, including the originating headquarters, subordinate commands, and separate intelligence support units.

(a) (U) Major Subordinate Commanders

(b) (U) Commanding Officer, Intel Bn

1 (U) OIC, IOC Support Cell

2 (U) OIC, IOC Surveillance and Reconnaissance Cell

3 (U) OIC, IOC Production and Analysis Cell

4 (U) Commanding Officer, CI/HUMINT Co.

5 (U) Officers in Charge, HUMINT Support Teams

(2) (U) Requests to Higher, Adjacent, and Cooperating Units. Provide separate numbered subparagraphs pertaining to each unit not organic, attached or supporting and from which CI support is requested, including other components, JTF headquarters, allied or coalition forces, theater and national operational and intelligence elements. Provide strengths, locations, capabilities, and type of support to be provided from external U.S. command and agencies and allied/coalition/host nation CI elements.

c. (U) Coordinating Instructions. Reference Appendix 16 (Intelligence Operations Plan), and command and other pertinent forces and organizations intelligence and CI SOPs. Detail here or in supporting tabs key changes to SOPs. Additional topics to include or emphasize here are: requesting CI support; direct liaison among subordinate commanders, MAGTF CI units, staff officers, and pertinent external organizations and agencies; routine and time-sensitive CI reporting procedures and formats, etc.

5. (U) Security. Provide planning guidance concerning procedures and responsibilities for the following security activities:

a. (U) Command Element and Other HQs

b. (U) Military Security

c. (U) Civil Authority

d. (U) Port, Border, and Travel Security

e. (U) Safeguarding Classified Information and Cryptographic Material Systems Resources

f. (U) Security Discipline and Security Education

g. (U) Protection of Critical Installations

h. (U) Special Weapons Security

i. (U) Counterterrorist Measures

6. (U) Counterintelligence Plans, Activities, and Functions

a. (U) Defensive. Identify the staff of those commands that have supporting CI assets and provide planning guidance concerning procedures, priorities, and channels for:

(1) (U) counterintelligence force protection source operations (CFSO)

(2) (U) Interrogation of EPW and defectors

(3) (U) Screening of indigenous refugees, displaced persons, and detained suspects

(4) (U) Debriefing of U.S. or other friendly personnel who evade, escape, or are released from enemy control

(5) (U) Exploitation of captured documents and material

b. (U) Offensive. Establish guidance, including control and coordination, for approval of counterespionage, countersabotage, countersubversion, counterterrorist, double agent, deception and other special operations.

7. (U) Counterintelligence Targets and Requirements.

a. (U) Targets. Reference Tab A (Intelligence Collection Plan) to Appendix 16 (Intelligence Operations Plan). Provide guidance for executing and managing CI collection activities not otherwise covered by regulation or SOP, equipment status, reports, and other specialized forms of collection activity to support the plan. Provide guidance on both routine and time-sensitive reporting of CI collected intelligence information by all CI collection sources to be employed in support of the plan. Provide guidance to MAGTF major subordinate commands/elements for developing CI targets based on an assessment of the overall CI threat. Designate priorities that emphasize the relative importance of the following CI target categories:

(1) (U) Personalities

(2) (U) Installations

(3) (U) Organizations and groups

(4) (U) Documents and material

b. (U) Priorities. Identify special CI collection requirements and priorities to be fulfilled by CI operations.

c. (U) Miscellaneous. Identify any other command information and intelligence required.

8. (U) Counterintelligence Production. Reference Tab B (Intelligence Production Plan) to Appendix 16 (Intelligence Operations Plan). Identify the CI production objectives and effort, including any intelligence and CI products required supporting the OPLAN. Include details of management of CI production requirements along with guidance on CI production and data bases, forms/formats for products, production schedules, CI products and reports distribution, etc. Address integration of CI analysis and production with all-source intelligence analysis and production activities. Include as appropriate requirements and guidance for the following: indications and warning, support to targeting, support to combat assessment (to include battle damage assessments), and especially CI support to force protection.

9. (U) Counterintelligence Dissemination. Reference Tab C (Intelligence Dissemination Plan), Tab D (Intelligence Communications and Information Systems Plan), and Tab E (Intelligence Reports) to Appendix 16 (Intelligence Operations Plan); and Annex K (Communications and Information Systems). Stipulate requirements, means and formats for disseminating CI reports and products (e.g., units responsible for each, periods covered, distribution, and timeline standards). Establish supporting CI communications and information systems plan and supporting procedures and criteria to satisfy expanded requirements for vertical and lateral dissemination of routine and time-sensitive CI products and reports. Address voice, network, courier, briefings, special CI communications, and other communications methods, including point-to-point and alarm methods. Establish alternate means to ensure that required CI will be provided to subordinate and supported units. Provide guidance regarding CI and information security, to include the dissemination of sensitive CI information within the force and the releasability of CI information and products to non-U.S. forces.

10. (U) Administration and Logistics. Provide a statement of the administrative and logistic arrangements or requirements for CI not covered in the basic pan or in another annex. Identify CI unique logistics and personnel requirements, concerns and

Page number

CLASSIFICATION

deficiencies. Discuss specific operational details on early deployments, mode of transportation, clothing, equipment, operational or contingency funds.

11. (U) Command and Control.

a. (U) Command and Control. Specify C2 command and support relationships and supporting information for all MAGTF CI elements. Include details of conditions that would prompt change of C2 relationships and procedures to implement that change during execution of the plan. Address what information and activities require the commander's knowledge and approval.

b. (U) CIS. Reference Appendix 16 (Intelligence Operations Plan) and Annex K (Communications and Information Systems). Ensure that CIS requirements are addressed in Annex K to the OPLAN or OPORD. Unique CIS requirement for CI operations should be addressed to include identifying what communication channels should be used for maintenance and administration of CI data bases, etc.

c. (U) Information Management. Provide any instructions necessary regarding information management (time-sensitive and routine reporting criteria, intelligence data bases, reports, etc.) that will influence MAGTF CI operations.

d. (U) Intelligence and CI C2 Nodes and Facilities. Reference the unit's intelligence SOP and Appendix 16 (Intelligence Operations Plan). Provide any guidance and instructions necessary regarding the establishment and operations of intelligence and CI C2 nodes and facilities (e.g., CI/HUMINT Co command post; CI representation within the surveillance and reconnaissance cell and the production and analysis cell, etc.).

e. (U) Coordination. Identify coordination requirements peculiar to CI activities listed in the paragraphs above.

f. (U) Reports. Identify CI reports that will be used and any necessary supporting information.

ACKNOWLEDGE RECEIPT

Name

Rank and Service

Title

TABS:

A - (U) Counterintelligence Estimate

B - (U) Counterintelligence List of Targets

C - (U) Countersigns Challenges and Passwords

Page number

CLASSIFICATION

This page intentionally left blank.

Counterintelligence Estimate

Purpose. Provides a baseline of historical, threat related CI information to support initial MAGTF.

CLASSIFICATION

Copy no. __ of __ copies
ISSUING HEADQUARTERS
PLACE OF ISSUE
Date/Time Group
Message reference number

TAB A TO APPENDIX 3 TO ANNEX B TO OPORD XXX (U)

COUNTERINTELLIGENCE ESTIMATE (U)

(U) REFERENCES:

(a) Unit SOP for intelligence and CI.

(b) JTF, NTF, other components, theater and national intelligence and CI plans, orders and tactics, techniques and procedures; and multinational agreements pertinent to intelligence operations.

(c) Maps, charts, and other intelligence and CI products required for an understanding of this annex.

(d) Documents and online data bases providing intelligence required for planning.

(e) Others as appropriate.

1. (U) Mission. (State concisely the CI mission as it relates to the MAGTF's planned operation.)

2. (U) Characteristics of the Area or Operations. (State conditions and other pertinent characteristics of the area that exist and may affect enemy intelligence, sabotage, subversive and terrorist capabilities and operations. Assess the estimated effects on friendly CI capabilities, operations, and measures. Reference appendix 11, Intelligence Estimate, to annex B, Intelligence, as appropriate.)

a. (U) Military Geography

(1) (U) Existing situation.

(2) (U) Estimated effects on enemy intelligence, sabotage, subversive and terrorist operations and capabilities.

(3) (U) Estimated effects on friendly CI operations, capabilities, and measures.

b. (U) Weather

(1) (U) Existing situation.

Page number

CLASSIFICATION

(2) (U) Estimated effects on enemy intelligence, sabotage, subversive and terrorist operations and capabilities.

(3) (U) Estimated effects on friendly CI operations, capabilities, and measures.

c. (U) Other Characteristics. (Additional pertinent characteristics are considered in separate subparagraphs: sociological, political, economic, psychological, and other factors. Other factors may include but are not limited to telecommunications material, transportation, manpower, hydrography, science, and technology. These are analyzed under the same headings as used for military geography and weather.)

3. (U) Intelligence, Sabotage, Subversive, and Terrorist Situation. (Discuss enemy intelligence, sabotage, subversive, and terrorist activities as to the current situation and recent/significant activities. Include known factors on enemy intelligence, sabotage, subversive, and terrorist organizations. Fact sheets containing pertinent information on each organization may be attached to the estimate or annexes, or may be consolidated in automated databases that can be accessed by MAGTF units—ensure those used are identified, and location/access information is provided.)

a. (U) Location and disposition.

b. (U) Composition.

c. (U) Strength, including local available strength, availability of replacements, efficiency of enemy intelligence, sabotage, subversive, and terrorist organizations.

d. (U) Recent and present significant intelligence, sabotage, and subversive activities/movements (including enemy knowledge of our intelligence and CI efforts).

e. (U) Operational, tactical, technical capabilities and equipment.

f. (U) Peculiarities and weaknesses.

g. (U) Other factors as appropriate.

4. (U) Intelligence, Sabotage, Subversive, and Terrorist Capabilities and Analysis. (List separately each indicated enemy intelligence, sabotage, subversive, and terrorist capability that can affect the accomplishment of the assigned MAGTF mission. Each enemy capability should contain information on what the enemy can do, where they can do it, when they can start it and get it done, and what strength they can devote to the task. Analyze each capability in light of the assigned mission, considering all applicable factors from paragraph 2, and attempt to determine and give reasons for the estimated probability of adoption by the enemy. Examine the enemy's capabilities by discussing the factors that favor or militate against its adoption by the enemy. The analysis of each capability should also include a discussion of enemy strengths and vulnerabilities associated with that capability. Also, the analysis should include a discussion of any indications that point to possible adoption of the capability. Finally, state the estimated

effect the enemy's adoption of each capability will have on the accomplishment of the friendly mission.)

a. (U) Capabilities

(1) (U) Intelligence. (Include all known/estimated enemy methods.)

(2) (U) Sabotage. (Include all possible agent/guerilla capabilities for military, political, and economic sabotage.)

(3) (U) Subversion. (Include propaganda, sedition, treason, disaffection, and threatened terrorist activities affecting our troops, allies, and local civilians, and assistance in the escape and evasion of hostile civilians.)

(4) (U) Terrorist. (Include capabilities of terrorist personalities and organizations in the AO.)

b. (U) Analysis and discussion of enemy capabilities for intelligence, sabotage, subversive, and terrorism as a basis to judge the probability of their adoption.

5. (U) Conclusions and Vulnerabilities. (Conclusions resulting from discussion in paragraph 4. Relate to current all-source intelligence estimates of the enemy's centers of gravity, critical and other vulnerabilities and estimated exploitability of these by friendly forces, enemy courses of action beginning with the most probable and continuing down the list in the estimated order of probability, and the estimated effects adoption of each capability would have on the friendly mission.)

a. (U) Probability of enemy adoption of intelligence, sabotage, subversive, and terrorist programs or procedures based on enemy's capabilities.

b. (U) Effects of the enemy's capabilities on friendly course of action.

c. (U) Effectiveness of our own CI measures and additional requirements or emphasis needed.

ACKNOWLEDGE RECEIPT

Name
Rank and Service
Title

EXHIBITS
(As appropriate)

Page number

This page intentionally left blank.

CLASSIFICATION

Copy no. ___ of ___ copies
ISSUING UNIT
PLACE OF ISSUE
Date/time group
Message reference number

TAB B TO APPENDIX 3 TO ANNEX B TO OPORD XXX (U)

COUNTERINTELLIGENCE LIST OF TARGETS (U)

1. (U) Friendly Infrastructure. Develop a listing of offices and agencies where CI personnel can obtain CI information and assistance.

2. (U) Foreign Intelligence and Security Service (FISS) Infrastructure. Develop a listing of specific offices and institutions within the FISS structure that can provide information of FISS targeting, operations, etc.

3. (U) FISS Personalities. Develop and update a specific listing of FISS personalities who, if captured, would be of CI interrogation interest.

ACKNOWLEDGE RECEIPT

Name
Rank and Service
Title

Page number

CLASSIFICATION

This page intentionally left blank.

CLASSIFICATION

Copy no. ___ of ___ copies
ISSUING UNIT
PLACE OF ISSUE
Date/time group
Message reference number

TAB C TO APPENDIX 3 TO ANNEX B TO OPORD XXX (U)

COUNTERSIGNS, CHALLENGES, AND PASSWORDS (U)

This tab provides the initial dissemination of the primary and alternate countersigns to be used within the MAGTF. Subsequent countersign dissemination will be made by other security means prior to the effective time.

Countersigns, Challenges, and Passwords

1. (U) Guidance and Procedures

a. (U) Countersigns (challenge/password) are used during MAGTF operations as a means of positive identification of friendly personnel. Countersigns will be changed daily at a predetermined time to be published in Annex C to the OPORD. Compromise of the countersign will be reported immediately to the MAGTF command element G-2/S-2 section.

b. (U) The countersigns list will be issued separately as Tab C to Appendix 3 to Annex B of the OPORD. It will appear in the following manner:

Table B-1.

Code	Challenge	Password	Alternate
11	Lamp	Wheel	9
12	Powder	Powder	7
13	Black	Table	8

c. (U) Dissemination of the initial primary and alternate countersigns for the initial introduction of forces will be made in Annex B to the OPORD. Subsequent countersign dissemination will be made by other secure means (i.e., covered radio nets) prior to the effective time.

A sample message form is as follows:

Code 11 countersign effective 011201(L) through 021200 (L). Alternate countersign Code 13.

Procedure: Alternate countersigns are any two numbers, that equal the alternate number, one given as the challenge, the other as the password reply.

Page number

CLASSIFICATION

d. (U) If at any time, there is reason to believe that a password or countersign has been compromised, the unit which suspects the compromise will notify the MAGTF command element G/S-2 via the fastest means available. The command element will issue alternate and any changes to the remaining countersigns.

e. (U) Below is the basic format for the countersigns, challenges, and passwords tab to appendix 3.

ACKNOWLEDGE RECEIPT

Name
Rank and Service
Title

APPENDIX C. COUNTERINTELLIGENCE PRODUCTION AND ANALYSIS

PART I. TACTICS, TECHNIQUES, AND PROCEDURES FOR C-HUMINT ANALYSIS AND PRODUCTION

Counter human intelligence (C-HUMINT) analysis increases in importance with each new U.S. involvement in worldwide operations. Especially in MOOTW, C-HUMINT analysis is rapidly becoming a cornerstone on which commanders base their concepts operations. This part presents information for analysts to develop some of those products that enhances the probability of successful operations.

CI analysts, interrogators, and CI agents maintain the C-HUMINT data base. Using this data base, they produce—

- Time event charts.
- Association matrices.
- Activities matrices.
- Link diagrams.
- HUMINT communication diagrams.
- HUMINT situation overlays.
- HUMINT-related portions of the threat assessment.
- CI target lists.

The analytical techniques used in HUMINT analysis enable analysts to visualize large amounts of data in graphic form. These analytical techniques are only tools used to arrive at a logical and correct solution to a complex problem; the techniques themselves are not the solution.

There are three basic techniques (tools) used as aids in analyzing HUMINT-related problems. Used together, these techniques—time event charting, matrix manipulation, and link diagramming—are critical to the process of transforming diverse and incomplete bits of seemingly unrelated data into an understandable overview of an exceedingly complex situation.

Time Event Charting

The time event chart (see figure C-1 on page C-2) is a chronological record of individual or group activities designed to store and display large amounts of information in compacted space. This tool is easy to prepare, understand, and use. Symbols used in time event charting are very simple. Analysts use triangles to show the beginning and end of the chart. They also use triangles within the chart to show shifts in method of operation or change in ideology. Rectangles or diamonds indicate significant events or activities.

Analysts can highlight particularly noteworthy or important events by drawing an X through the event symbol (rectangle or diamond). Each of these symbols contain a chronological number (event number), date (day, month, and year of event); and possibly a file reference number. The incident description is a brief explanation of the incident, and may include team size, type of incident or activity, place and method of operation, and duration of incident. Time flow is indicated by arrows.

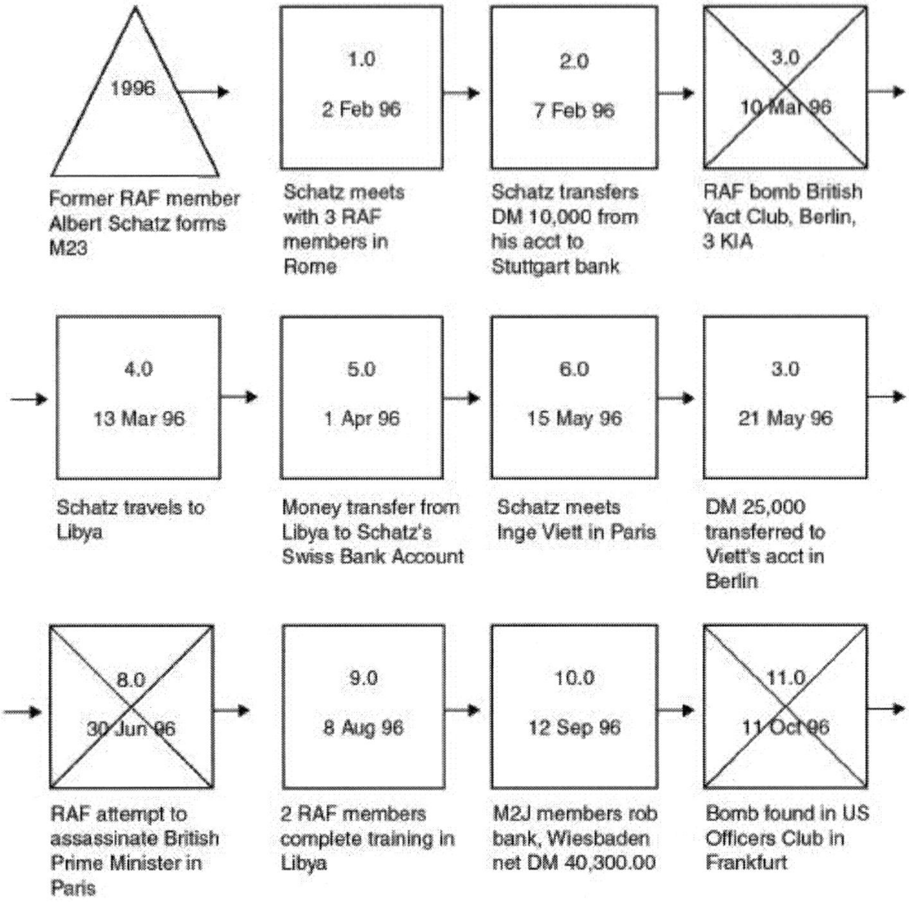

Figure C-1. Time Event Chart.

Analysts also use a variety of symbols such, as parallelograms and pentagons to show different types of events and activities. Using these symbols and brief descriptions, the CI analyst can analyze the group's activities, transitions, trends, and operational patterns. Time event charts are both excellent briefing aids and flexible analytical tools.

Matrix Manipulation

A matrix is the optimum way to show relationships between similar or dissimilar associated items. Items can be anything relevant to the investigation: persons, events, addressees, organizations or telephone numbers. Analysts use matrices to determine who knows whom or who has been where or done what. This results in a clear and concise display that viewers can understand easily by looking at the matrix.

Matrices resemble the mileage charts commonly found in a road atlas. There are two types of matrices used in investigative analysis: the association matrix and the activities matrix.

Association Matrix

The association matrix shows an existing relationship between individuals. In HUMINT analysis, the part of the problem deserving the most analytical effort is the group itself. Analysts examine the group's members and relationships with other members, and related events. Analysts can show connections between key players in any event or activity in an association matrix (see figure C-2). It shows associations in a group or similar activity and is based on the assumption that people involved in a collective activity know each other.

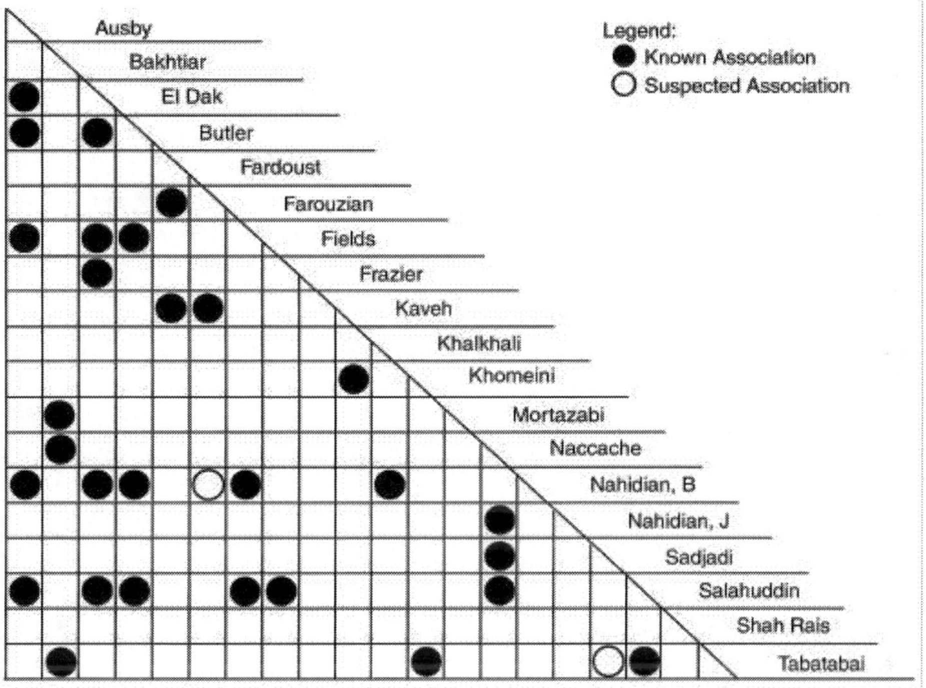

Figure C-2. Association Matrix.

This type of matrix is constructed in the form of a right triangle having the same number of rows and columns. Analysts list personalities in exactly the same order along both the rows and columns ensuring possible associations are shown correctly. The personality matrix shows who knows whom. Analysts determine a known association by direct contact between individuals. They determine direct contact by a number of factors; face-to-face meetings, confirmed telephonic conversation between known parties, and members of a particular organizational cell.

Note: When a person of interest dies, a diamond is drawn next to the person's name on the matrix.

CI analysts indicate a known association between individuals on the matrix by a dot or filled-in circle. They consider suspected or weak associations between persons of interest to be possible or even probable, but cannot be confirmed using the previous criteria.

Examples of suspected associations include—

- When a known party calls a known telephone number (analysts know to whom the telephone number is listed), but cannot determine with certainty who answered the call.
- When an analyst can identify one party to a face-to-face meeting, but may be able to only tentatively identify the other party.

Weak or suspected associations on the matrix are indicated by an open circle. The rationale for depicting suspected associations is getting as close as possible to an objective analytic solution while staying as close as possible to known or confirmed facts. If analysts confirm a suspected association, they can make the appropriate adjustment on the personality matrix.

A secondary reason for depicting suspected associations is that it gives analysts a focus for tasking limited intelligence collection assets to confirm suspected associations.

Note: The association matrix: it shows only that relationships exist; not the nature, degree, or frequency of those relationships.

Activities Matrix

The activities matrix determines connectivity between individuals and any organization, event, entity, address, activity or anything other than persons. Unlike the association matrix, the activities matrix is constructed in the form of a square or a rectangle (see figure C-3). It does not necessarily have the same number of rows and can tailor rows or columns to fit current or future requirements. The analyst determines the number of rows and columns by needs and the amount of information available.

Analysts normally construct this matrix with personalities arranged in a vertical listing on the left side, and activities, organizations, events, addresses or any other common denominator arranged along the bottom.

This matrix stores an incredible amount of information about a particular organization or group, and builds on information developed in the association matrix. Starting with fragmentary information, the activities matrix reveals an organization's—

- Membership.
- Organizational structure.
- Cell structures and size.
- Communications network.
- Support structure.
- Linkages with other organizations and entities.
- Group activities and operations.
- Organizational and national or international ties.

As with the association matrix, known association between persons and entities is indicated by a solid circle, and suspected associations by an open circle.

Analysts use matrices to present briefings, present evidence, or store information in a concise and understandable manner within a data base. Matrices augment, but cannot replace, standard reporting procedures or standard data base files.

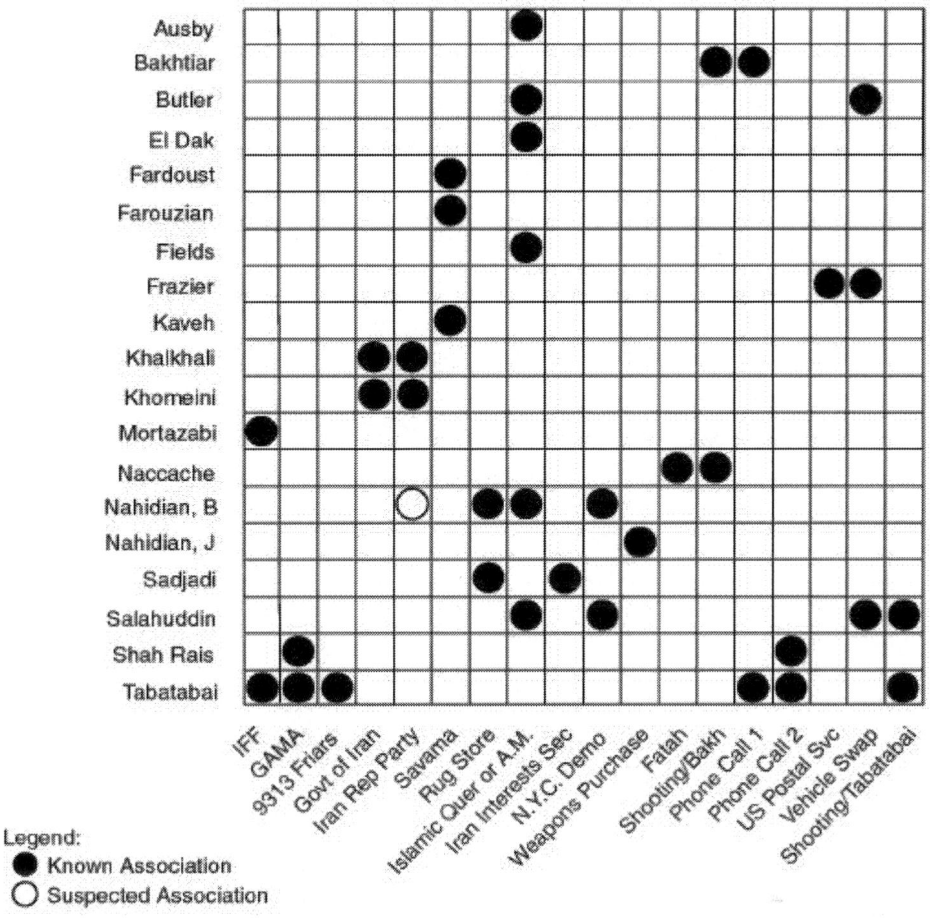

Figure C-3. Activities Matrix.

Using matrices, the analysts can—

- Pinpoint the optimal targets for further intelligence collection.
- Identify key personalities within an organization.
- Increase understanding of an organization and its structure.

Link Diagramming

The third analytical technique is the link diagram (see figure C-4 on page C-6). Analysts use this technique to depict the more complex linkages between a large number of entities, such as persons, events or organizations. Analysts use link analysis in a variety of complex investigative efforts including criminal investigations, terrorism, analysis, and even medical research. Several regional law enforcement training centers are currently teaching this method as a technique in combating organized crime. The particular method discussed here is an adaptation useful in general CI investigative analysis, particularly terrorism.

The difference between matrices and link analysis is the same as the difference between a mileage chart and a road map. The mileage chart shows the connections between

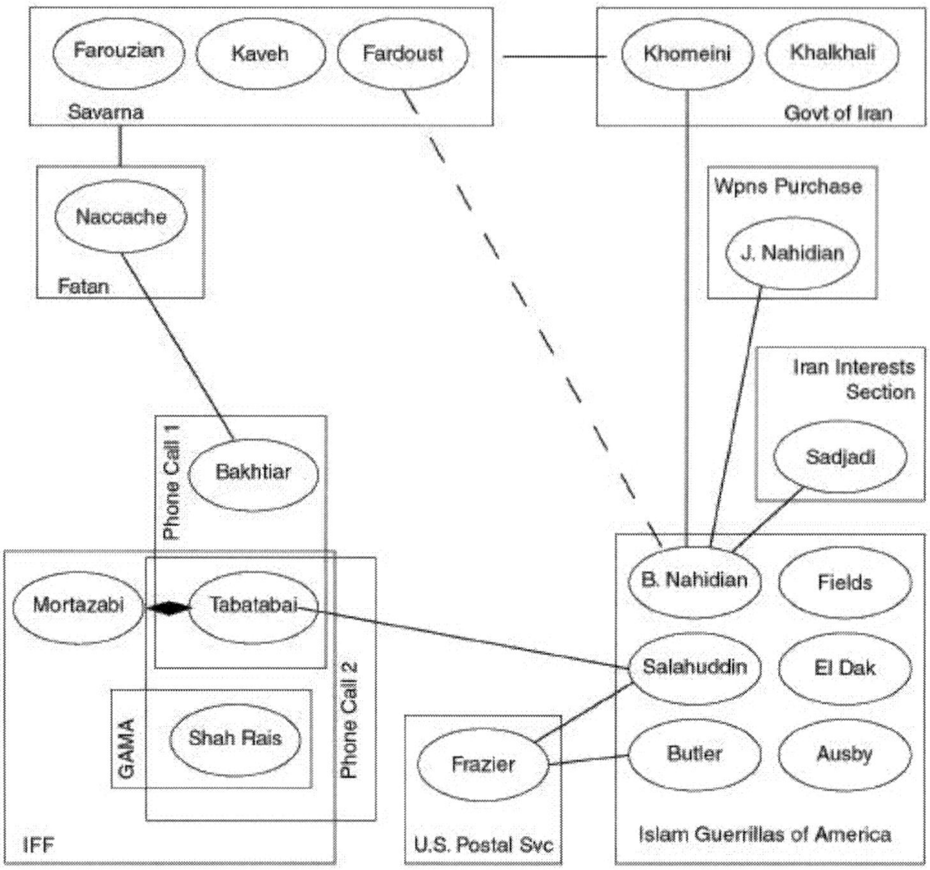

Figure C-4. Link Diagram.

cities using numbers to represent travel distances. The map uses symbols that represent cities, locations, and roads to show how two or more locations are linked to each other. Different symbols on the map have different meanings, and it is easy to display or discover the best route between two or more locations as well as identify obstacles such as unpaved roads or bodies of water.

The same is true with link analysis. Different symbols identify different items. Analysts can easily and clearly display obstacles, indirect routes or connections, and suspected connections. Often, the viewer can work with and follow the picture easier than the matrix. Link analysis presents information that ensures clarity.

As with construction of association matrices, certain rules of graphics, symbology, and construction must be followed. Standardization is critical to ensure those who construct, use or read a link diagram understand exactly what the diagram depicts.

- Persons—open circles with names written inside the circle.
- Persons known by more than one name (alias, also known as [AKA])— overlapping circles with names in each circle.
- Deceased persons same as previous, but with a diamond next to the circle representing that person.

- Non-personal entities (organizations, governments, events, locations)— squares or rectangles.
- Linkages or associations—solid lines for confirmed and dotted lines for suspected.

Note: Each person or non-personal entity is shown only once in a link diagram.

Certain conventions must be followed. For clarity, analysts arrange circles and squares so that whenever possible, lines of connectivity do not cross. When dealing with a large or especially complex problem, it is difficult to construct a link diagram so that no connecting lines cross. Intersecting lines muddle the drawing and reduce clarity. If lines must cross, show the crossing in exactly the same manner as on an electrical schematic or diagram.

Link diagrams can show organizations, membership within the organization, action teams or cells, or participants in an event. Since each individual depicted on a link diagram is shown only once, and some individuals may belong to more than one organization or take part in more than one event, squares or rectangles representing non-personal entities may overlap.

Construct the appropriate association matrices showing who knows whom, who participated in what, who went where, and who belongs to what group.

Draw information from the data base and intelligence reports, and relationships from the matrices. Group persons into organizations or cells based on information about joint association, activities or membership. Draw lines representing connections between individuals, organizations or activities to complete the diagram. The diagram may require rearrangement to comply with procedural guidelines, such as crossed lines of connectivity. The finished product will clearly display linkages between individuals, organizations, and other groupings.

When the matrices and link diagram are complete, recommendations are made about the group's structure and areas identified. Identify areas for further intelligence collection targeting. Task intelligence assets to confirm suspected linkages and identify key personalities for exploitation or neutralization. The combination of matrix manipulation and the link diagram present a clear and concise graphic depiction of an extremely complex threat situation.

There is more to overlapping organizations than is immediately obvious. At first glance, the overlap indicates an individual may belong to more than one organization or has taken part in multiple activities. Further study and analysis may reveal connections between organizations, events, or organizations and events. When an organization or incident shown in a link diagram contains the names of more than one individual, it is unnecessary to draw a solid line between those individuals to indicate connectivity. It is assumed individual members of the same group or participants in the same activity know each other, and the connection between them is therefore implied.

A final set of rules for link diagrams concerns connectivity between individuals who are not members of an organization or participants in an activity, but who are somehow connected to the group or activity. Two possibilities exist: The individual knows a member or members of the organization but is not directly connected with the organization itself. The person is somehow connected with the organization or activity but cannot be directly linked with any particular member of that organization or activity. In the first case, the connectivity line is drawn between the circle representing the individual and the circle representing the person within the organization or activity.

PART II. TACTICS, TECHNIQUES, AND PROCEDURES FOR COUNTER-IMAGERY INTELLIGENCE PRODUCTION AND ANALYSIS

The proliferation of imagery systems worldwide, especially the platforms carrying imagery systems, complicates the task of C-IMINT analysts. Relatively inexpensive platforms that are easily transported and operated, such as unmanned aerial vehicles, are becoming available to anyone who wants to employ them. For the more sophisticated, there are other platforms either continuously circling the planet or in geosynchronous orbit, available for hire by anyone with the desire and the ability to pay the freight. An adversary need not possess the technology to build and launch such a platform. Adversaries merely buy time from the operators of the platform and obtain the products acquired during their allotted time. Like other CI functions, C-IMINT depends on analysts knowing the adversary and knowing ourselves. It begins long before friendly forces deploy for any operation and continues throughout the operation. It goes on even after our forces return to their home station after completion of the operation. C-IMINT begins with knowledge. CI analysts must have a thorough knowledge of the threat in the objective area and any threat from outside the AO that may influence our operations.

Predeployment

Prior to any operation, CI analysts need to prepare in-depth. In addition to researching data on the threat and the AO, analysts gather information and build a data base to serve C-IMINT in the coming operation. During this phase, analysts initiate quick reference matrices and the IMINT situation overlay.

Adversary Intelligence Flight Matrix

These matrices are concerned with other platforms used by the adversary. Tracking these collection systems continuously allows analysts to analyze threat IMINT collection patterns.

System Component Quick Reference Matrix

These matrices are concerned with adversary system's capabilities and processing times (see table C-1). This file is part of the data base that equates to an OOB file on threat IMINT systems.

IMINT Situation Overlays

These are the paths of adversary intelligence collection flights depicted on the friendly operations graphics. They identify areas susceptible to collection.

Friendly Patterns

Pattern analysis is the detailed study of friendly activities to determine if a unit performs the activities in a predictable manner, thus creating a monitorable pattern of activity. These actions cue an observer to a unit's type, disposition, activity, and capability. Imagery coverage of the AO is essential for planning and for reference later

Table C-1. System Component Quick Reference Matrix.

System Component Quick-Reference Matrix					
System: ______________			Date: ______________		
Organization	Location	Characteristics	Strength	Tactics	Remarks

during operations. Small or intermediate scale imagery covering the entire AO may be obtained from general reference files or national sources and need not be newly collected. The presence of U.S. reconnaissance aircraft making numerous passes over territory belonging to another nation would tip off an impending operation. File imagery or imagery obtained by satellite may be the only reference available.

When available and of high enough priority friendly IMINT is used to determine friendly patterns that may be susceptible to IMINT collection. These patterns are key indicators to the enemy of specific operational activities. Patterns usually occur because of a unit's SOP and doctrine. Example patterns include—

- Relocating fire support units forward before an attack.
- Locating command posts and other C2 facilities in the same relative position to maneuver elements and to each other.
- Repeating reconnaissance overflights of areas planned for ground or air attack about the same time before each operation.

Information gained from imagery provides a means of checking other reports and often produces additional detailed information on a specific airborne interceptor. Friendly activities thus need to be examined collaterally with imagery of a particular area. Imagery provides confirmation of installations, lines of communications, and operational zones. Side looking airborne radar (SLAR), for example, detects night movements of watercraft.

In the overall evaluation, analysts synthesize the separate trends developed during analysis. Such a process identifies the possible compromise of an existing element, activity or characteristic based on logical relationships and hypotheses developed by analysis. The pattern analysis technique is just one of many techniques designed to help evaluate friendly units for vulnerability to threat IMINT. The process is a continuous one.

Analysis of a unit's movements gives significant clues to its intentions, capabilities, and objectives. By applying this technique against our own units, analysts identify vulnerabilities. Movement analysis forms an important step in the identification and recommendation of countermeasures.

SLAR is a primary sensor in detecting moving targets or moving target indicators and is usually associated with the special electronics mission aircraft and joint surveillance target attack radar system platform. While the sensor is primarily focused at enemy moving target indicators, it identifies friendly movement patterns that may also be collected by the enemy.

Tracks created by a unit give excellent indication of a unit's disposition. Any time a unit moves away from hard packed roads, the danger of leaving track signatures is high. The following countermeasures should be observed to disguise or eliminate these signatures:

- Conceal tracks by netting or other garnish.
- Disperse turnouts near command posts.
- Place installations and equipment near hard roads where concealment is available.

Using our IMINT resources helps determine the effectiveness of a friendly unit's program to suppress its visual and thermal signatures, including positioning of assets. Friendly aerial reconnaissance is extremely limited and must be planned well in advance. The following are examples of countermeasures used to reduce our vulnerability to enemy IMINT:

- Use traffic discipline when moving into and out of the installation; this may require walking some distance to a CP.
- Drive in the treelines when roads are not available.
- Extend new roads beyond the CP to another termination.
- Control unauthorized photographic equipment.
- Use physical security measures to prevent optical penetration.
- Use proper camouflage procedures.
- Limit the dissemination of photographs made within the installation.
- Avoid use of direction signals and other devices that provide information.
- Conceal equipment markings.
- Prevent detection by infrared imaging (nets, infrared generators).
- Eliminate open-air storage of special equipment, raw materials, and telltale objects.

The key to proper positioning of assets on the ground is to use natural features as much as possible. Obvious locations such as clearings may be more convenient but should be avoided. Infrared and SLAR missions are particularly effective at night. Units should be well dispersed since a high concentration of tents and vehicles, even well hidden, will stand out on imagery to trained analysts.

Evaluation of Countermeasures

For these countermeasures to be effective, every command should develop a self-evaluation system to ensure proper employment.

PART III. TACTICS, TECHNIQUES, AND PROCEDURES FOR COUNTER-SIGNALS INTELLIGENCE PRODUCTION AND ANALYSIS

Threat SIGINT Capabilities and Assessment

One of the key words in the definition of intelligence is enemy. We need to know and understand the capabilities and limitations of the threat arrayed against us and how the threat can influence our operations and mission. The first step in the C-SIGINT process provides extensive information on determining foreign technical and operational capabilities and intentions to detect, exploit, impair, or subvert the friendly communications and electronic environment.

Threat assessment is the key in planning C-SIGINT operations. Subsequent steps are necessary only when a defined threat exists.

Threat assessment is a continuous activity. It takes place throughout the conflict spectrum. A specific threat assessment is required to support a specific operation or activity.

CI analysts gather and analyze information. They interact with staff elements and higher, lower, and adjacent units to obtain the necessary data and access to supportive data bases. Command support and direction are essential to success in the threat assessment process.

Major information sources available to analysts include—

- Validated finished intelligence products.
- Theater and national level SIGINT threat data base.
- Previous tasking.
- Analyst experience.
- The CI data base.

CI analysts must continue to refine this list and identify other sources of information available for their particular AO.

There are six tasks associated with the C-SIGINT threat assessment (see figure C-5 on page C-12).

Note: Within a MAGTF, C-SIGINT production and analysis generally results from the integrated operations of the P&A cell, the radio battalion OCAC, and the supporting CI/HUMINT company CP.

Task 1—Identify Threat Systems in the Geographic area of Responsibility. This task provides the initial focus for the remaining threat assessment tasks. The primary objective of this task is to determine the specific threat faced by the MAGTF. Analysts collect required data to properly identify the threat. Additionally, analysts must coordinate and request assistance from the collection management element. The procedures for identifying the threat systems follow:

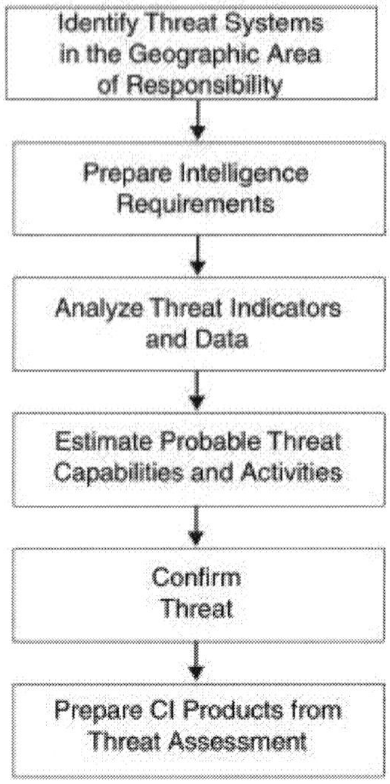

Figure C-5. SIGINT Threat Assessment Process.

Identify the generic threat. Analysts enter the CI data base and retrieve the most recent appropriate threat assessment. Analysts then review this data to determine what threat systems were known to be in their AO on the date of the assessment. Next, analysts examine finished intelligence products published by national level agencies to obtain technical and operational data on the threat system. Some of the intelligence products include—

- Electronic Support and Electronic Attack capability studies.
- SIGINT threat by country.
- SIGINT support to combat operations.

Create the doctrinal template. The doctrinal template is a graphic display of threat's systems deployment when not constrained by weather and terrain. The analyst should review the database for existing templates before constructing a new one.

Collect data. Data collection is required when analysts receive tasking for a specific unit or operation. Analysts must collect additional data to identify the threat to a particular unit or AO.

Create the SIGINT situation overlay. Analysts review the collected data to determine—

- Technical and operational capabilities.
- Typical modes of operation.
- Current deployment.
- Probable tasking.
- Activities of the collectors of interest.

Enter data. Analysts enter this data on the situation overlay.

Summarize the data and identify the threat system. CI analysts review the SIGINT situation overlay for patterns, electronic configurations, and threat C2, CIS and EW. Generally this information is available from either intel bn's P&A cell or the radio battalion's OCAC. A common approach is to pose and answer questions, such as—

- Is the threat system part of a larger system?
- What are the threat system's capabilities?
- How is the threat system doctrinally used?
- How does the threat system obtain information?
- How many collection systems were located?

Request Intelligence. In some instances, sufficient information may not be available in the unit to make an accurate determination. For example, the type of equipment may be known but the technical characteristics of the system may not be available from local sources. If additional intelligence is required, CI analysts compile the information needed and coordinate with the MAGTF collections manager to request additional intelligence from outside the unit.

Task 2—Prepare Information Requirements

CI analysts fill information shortfalls by requesting information from sources external to the unit. These external information sources are adjacent or higher echelons and national level assets. Each echelon satisfies a request with available data or organic assets, if possible. Requirements exceeding their organic capabilities are consolidated and forwarded to the next higher echelon as a request for information.

Task 3—Analyze Threat Indicators and Data

CI analysts review, organize, and evaluate key information components of the collected information. They update the data looking for trends and patterns of the threat system that provide an estimate of capabilities and intentions. They focus on each component of the collected information to determine if it reveals a tendency of the threat system to act or react in a particular manner. Additionally, analysts evaluate the information for trends or characteristics that aid in the ID and evaluation of the capabilities and intentions of the threat system. Additional support may be required from other staff elements. Procedures for analyzing threat indicators and data are to—

Compile and organize data. First, analysts compile and organize the data that has been collected. They update the data base with new information and organize the data into collector categories.

Review data. Analysts review the collected data to determine the ability of the threat systems to collect against a specific target.

Determine intentions. To determine the intentions of the threat system, CI analysts pose the following questions and enter this information in the data base:

- What area will the threat system target?
- When will the targeting take place?
- Why is the targeting taking place?
- How will the threat system attempt to collect against the target?

- How has the threat system been used in the past?
- What does threat doctrine suggest about probable threat?
- Does the threat system have a distinctive signature?

Doctrinal templates are extracted from the data base and compared to the SIGINT situation overlay. Analysts list similarities between current and doctrinal deployments and select the doctrinal template that has the greatest similarity to the current situation.

Task 4—Estimate Probable Threat

CI analysts identify the probable threat. They review the information collected and apply this information to the geographic AOI and the capabilities and intentions of the threat system. Procedures for predicting the probable threat follow:

Determine probable location. Use the SIGINT situation overlay and doctrinal templates to determine the location of the collectors. Overlay the doctrinal template onto the situation overlay.

Analyze terrain and weather effects. Integrate the terrain and weather data with the doctrinal template and the SIGINT situation overlay and create a situation template for the current environment. Terrain and weather conditions affect a threat system's ability to operate according to their doctrine. For example, a radio direction finding site must have a clear line of sight on the emission of the target to gain an accurate bearing. Mountains, dense foliage, and water distort electronic emissions and impair a collector's ability to target.

Update the SIGINT situation overlay. Place the symbols for the collectors on the doctrinal template that have not been confirmed on the SIGINT situation overlay as proposed locations.

Task 5—Confirm Threat

CI analysts attempt to verify threat predictions. The procedures for confirming the threat follow.

Validate existing data. Review current intelligence reports and assessments to determine if the information received in response to requests for intelligence in the assessment are valid. If there are indications that the capabilities or intentions of the threat system have changed, additional information may be required. This is determined by looking for information that could indicate a change in a collector's ability to collect against the command. For example, additional antennas added to the collector, or the collector moved to provide for better targeting indicating a change in collection capabilities.

Request additional information. If additional information is required, these intelligence requirements will be tasked to organic intelligence units or submitted to higher headquarters.

Evaluate new information. If new information on the collector's intentions or capabilities is received, review this information to determine its impact on the original assessment, and update the situation overlay. If intentions and capabilities of the collector change, reevaluate the original threat prediction by following the tasks identified in previous sections.

Task 6—Prepare CI Products from SIGINT Threat Assessment

CI analysts can present the SIGINT threat assessment in briefings or reports. Portions of the threat assessment are included and presented in other CI and all-source intelligence products.

MAGTF Vulnerability Assessment

After examining the enemy's SIGINT and EW equipment, capabilities, and limitations, our own unit must be examined to see how our adversary can affect us. The second step in the C-SIGINT process details specific areas where a threat effort can be most damaging to the friendly force.

Vulnerabilities are ranked according to the severity of their impact on the success of the friendly operation. The vulnerability assessment—

- Examines the command's technical and operational communications-electronics (C-E) characteristics.
- Collects and analyzes data to identify vulnerabilities.
- Evaluates vulnerabilities in the context of the assessed threat.

CI analysts perform the primary data gathering and analysis required. Assistance by appropriate staff elements (intelligence, operations, CIS) is key to this process.

Data gathering requires access to command personnel and to local data bases. Data sources include—

- Technical data on C-E inventories.
- Doctrinal and SOP information.
- Output from the threat assessment step.
- Command friendly force information.
- EEFI.
- PIRs and IRs.

The data base of friendly technical data is used throughout the vulnerability assessment process for key equipment information, mission data, and other supporting information.

MAGTF vulnerability assessment is comprised of ten tasks. The first three tasks are ongoing determinations of general susceptibilities. The next six are specific to the commander's guidance and involve determinations of specific vulnerabilities. The final task is the output. MAGTF vulnerability assessment tasks are shown in figure C-6 on page C-16.

Task 1—Compile Friendly C-E Characteristics

CI analysts compile friendly C-E characteristics. They collect and organize unit C-E data and equipment characteristics for analysis. This analysis provides a baseline for analyzing friendly C-E equipment and operational susceptibilities to threat operations. The compilation of C-E characteristics is an ongoing process. Assistance from the CIS and EW officers provide needed information.

The C-E data are a baseline for identifying friendly susceptibilities. A unit's equipment, personnel, and associated characteristics must be identified before the pattern and

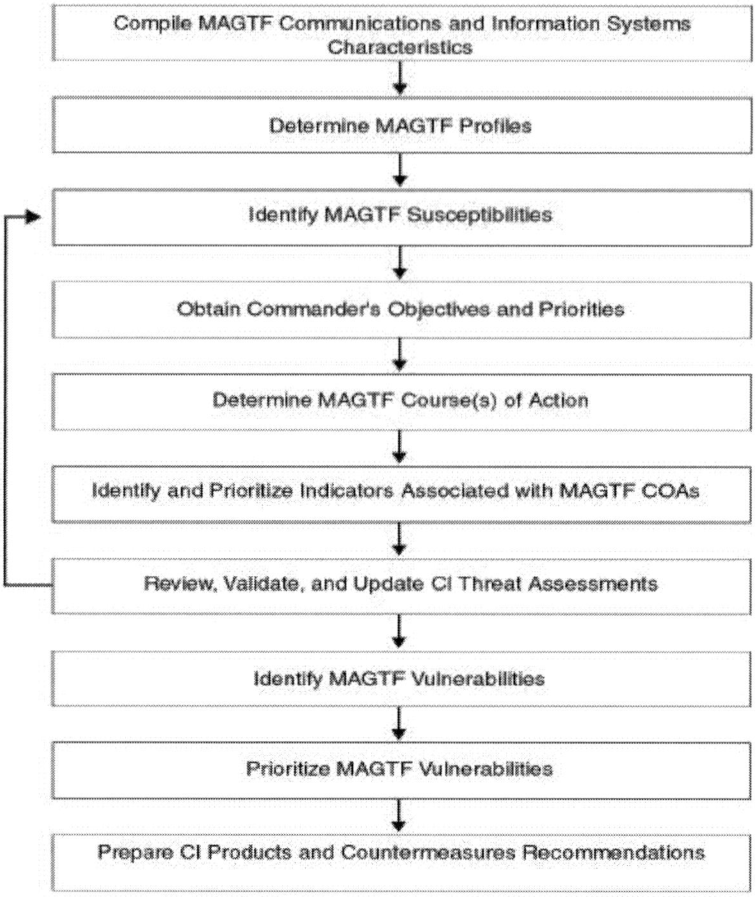

Figure C-6. MAGTF Vulnerability Assessment Process.

signature analysis can proceed. The CI analyst uses available databases to extract the table of equipment (T/E) and technical manuals (TM) on MAGTF C-E equipment. The following are procedures for compiling friendly C-E characteristics.

Gather data on friendly C-E characteristics. Gather C-E data and characteristics of the equipment. Identify the following types of C-E data:

- T/Es, TMs, and technical data for C-E equipment in a unit.
- References describing the unit and its equipment configuration.
- Current maintenance levels and normal status of the equipment.
- Personnel status, including current training levels of personnel in the unit.
- Equipment performance capabilities and operational capabilities in all weather conditions, at night, over particular terrain, and toward the end of equipment maintenance schedules.
- Equipment supply requirements.
- Special combat support requirements.

Organize C-E data. CI analysts organize the information into a format useful for signature analysis. The data are organized by type of unit (if the support is multi-unit), type of emitter, frequency range, number and type of vehicles or weapons that emit or carry emitters and type of cluster.

Task 2—Determine Friendly Force Profiles.

This task includes the analysis of signatures and patterns of the C-E equipment and a summary statement of the unit's C-E profile. A profile consists of the elements and standard actions, equipment, and details of a unit, the sum of signatures and patterns.

SIGNATURES + PATTERNS = PROFILE

Procedures for determining the friendly force's profile follow.

Analyze friendly force signatures. CI analysts—

- Extracorganic equipment characteristics for the operation.
- Determine environmental effects.
- Determine C-E characteristics for each friendly COA.
- Determine C-E equipment employment.
- Compare planned use with technical parameters.
- Determine if further evaluation is required.
- Perform tests with support from unit or higher echelon assets.
- Evaluate the information collected above.
- Diagram physical and electronic signatures on an overlay or other product.
- Update the CI data base.

Perform friendly pattern analysis. Identify standard practices, common uses of a unit's C-E equipment, and operational patterns by—

- Reviewing the data base to obtain information that might provide the threat with critical data regarding unit type, disposition, activities, or capabilities.
- Extracting from the OPLAN and OPORD particular means of communication, operational characteristics, and key and secondary nodes for communications support.
- Identifying specific patterns associated with types of operations.

Correlate patterns and signature. In this subtask, compile the information from the signature and pattern analysis, which creates the profile. Analysts—

- List the signature and pattern data for particular types of C-E equipment.
- Match signature with patterns to form the profile.
- Organize data into types of C-E operations.
- Correlate signature and pattern data with past profiles to produce the current profile shown in table C-2 on page C-18.

Produce unit profile. Patterns and signatures can change as commanders, staff, and operators change. Profile development must be an ongoing effort. To produce the unit profile, use the OPORD to obtain the past task organization and then select the areas of concern to that organization; C2, intelligence, maneuver, fires, logistics, and force protection.

Table C-2. Friendly Unit C-E Profile.

Command and Control		
Physical Signatures	**Electronic Signatures**	**Pattern Data**
• Types of vehicles • Number of vehicles • Distances to subordinate, adjacent, and higher headquarters	• Types of emitters • Frequency range • Signature type and range • Emitter fingerprints	• Timing of movement • Mode of movement • Collected or nearby units • Frequency of redeployments • Radio and radar net employment
Operations and Maneuver		
Physical Signatures	**Electronic Signatures**	**Pattern Data**
• Types of vehicles • Number of vehicles • Distances to subordinate, adjacent, and higher headquarters • Types of weapon systems	• Types of emitters • Frequency range • Signature type and range • Emitter fingerprints	• Timing of reconnaissance • Mode of reconnaissance • Timing of movement • Type of movement • Mode of movement • Units involved • Mode and source of supply

Task 3—Identify Friendly Susceptibilities

Analysts determine how the profiles would appear to threat systems and which equipment or operations are susceptible. A susceptibility is defined as the degree to which a device, equipment, or weapon system is open to effective attack due to one or more inherent weaknesses. Any susceptibilities are potential vulnerabilities.

Information Sources are of the following types:

- Current friendly C-E profile.
- Historical profiles to compare with current profile.
- Knowledge and experience from other analysts.

The procedures for identifying susceptibilities follow:

- Identify weaknesses:
  - Review current profile and identify unique equipment or characteristics that the threat may use to determine intentions.
  - Review the CI data base and compare historical profiles with current profile, noting correlations and deviations.
  - Plot friendly weaknesses to threat operations on a MAGTF electronic order of battle overlay.
- Categorize susceptibilities. Categorize susceptibilities to allow more specific analysis by equipment type, organization, and use. Do this—
  - By type (for example, equipment, operations, or both).
  - By activity (for example, logistic, CIS, intelligence, operations, and fire support).
  - According to resource requirements.
  - According to the length of time the susceptibility has existed.
  - According to scope (number of units or equipment types).

Task 4—Obtain Commander's Operational Objectives and Guidance
Commanders state operational objectives for missions in OPLANs and OPORDs. Analysts use this information to plan the most effective support for the commander and to identify the commander's preferences for types of operations. The commander's operational concept and the following example of a unit EEFI statement are essential to the analysis of friendly COAs.

Friendly Supported Unit: 1st MARDIV

1. Subordinate Element: HQ, 1st MARDIV
2. Location: 32U NB51452035
3. Tactical Objectives(s): Defend to PL Gray
4. Essential Elements of Friendly Information:
   a. Significant Compromises:
      (1) Time of counterattack.
      (2) Identification and location of battalions and higher headquarters elements.
      (3) Identification of attached units.
      (4) Loss/degradation of main C2 centers or supporting communications and information systems.
   b. Insignificant Compromises: identification of 1st MARDIV.

This information enables analysts to evaluate indicators of friendly COA in the context of what the commander considers essential to the success of the operation. Setting priorities for the vulnerabilities depends on the commander's operational concept. The primary information sources are:

- Concept of operation.
- OPORDs.
- OPLANs.
- Prioritized EEFIs.

Task 5—Determine Friendly COAs.
Based on the general description of the commander's objectives, the operations element plans locations and events. Analysts produce an overlay of the friendly force profile integrated with the commander's objectives. The procedures for determining friendly COAs follow.

Identify COA. For each applicable level of command, identify friendly COAs. At division level, for-example, COAs include the following minimum information:

- Summary of operations.
- Higher headquarters support.

Compare COA to Specific EEFIs. Review each COA for events or actions that could compromise the unit's mission by disclosing key EEFI. The review is summarized in an events list that describes a particular mission, COA, or event that may compromise the EEFI or the friendly intentions.

Task 6—Determine Indicators of Friendly COAs

Indicators of friendly COAs are those events and activities that, if known by the threat, would compromise a friendly COA. The procedures for determining indicators of a friendly COA follow:

Identify the Commander's Preferences and Perceptions about C-SIGINT Operations. Seek information about the commander's style from sources such as previous concepts, plans, and orders, or interviews with subordinate commanders and staff officers.

Integrate Friendly Profiles and COA. If planned location or movement data are not available, retrieve friendly operational overlays shown from the data base. Overlays help identify friendly historical positions for the new COA. Then integrate friendly force profiles and COAs:

- Noting current position and expected COA.
- Identifying key C-E capabilities associated with the COA (for example, radio nets, types of radios, radar, and teletypewriters).
- Noting past C-E operational patterns.
- Plotting critical C-E nodes, paths, or circuits.

Determine standard C-E procedures for Types of Operations.

- Use the commander's objectives to identify key operational constraints; nodes, paths, chokepoints, and standard C-E procedures followed during a particular COA. New or critical data, not previously included in the friendly profile and COA integration, are then added to the situation overlay.
- Consider constraints and procedures while determining indicators. Document these as factors associated with those indicators. After completing the review of existing data as obtained from the commander's objectives, determine what additional information is required.

Determine Impact of Weather and Terrain. As the situation changes, the significance of particular nodes or paths may shift or additional nodes may become critical. Consider the following in determining the impact:

- Inclement weather.
- Night activity.
- Terrain masking.
- Poor C-E equipment maintenance.

Set priorities. Once the type of operation is determined, set priorities for the events, movements, and nodes by their overall importance to the operation.

Identify Critical C-E Nodes.

- Determine key indicators of friendly operations using the C-E constraints and procedures identified from the information provided by the commander and data obtained from previous tasks. For each COA, extract those preparations, activities, or operations that could tip off the threat to the particular COA.
- List the indicators associated with a COA. Any special factors such as operational constraints, optimum weather conditions, or terrain requirements associated with an indicator should be described accordingly.

Task 7—Review and Validate Threat Assessment Data

Threat assessment data are further refined to proceed with the remainder of the vulnerability assessment. Analysts organize threat data in a format comparable to the friendly forces data. Missing data is identified and requested. The C-SIGINT analyst performs the review and validation of threat data with considerable exchanges of information with other analysts. The procedures for reviewing and validating threat assessment data follow.

Summarize and Reorganize Threat Assessment Data.

- Compile recent threat assessment information.
- Identify information shortfalls.
- Coordinate with the collection management section to initiate requests for information.

Extract Relevant Data for Vulnerability Assessment.

- Extract areas of threat operations most critical to the supported command.
- Document threat capabilities and intentions.
- Store data for later application.

Task 8—Identify Friendly Vulnerabilities

Analysts compare the enemy's intelligence collection threat with the friendly unit susceptibilities to determine the vulnerabilities. Once the vulnerabilities have been identified, analysts can rank them. The procedures for identifying vulnerabilities follow:

Compare Current Threat to Friendly C-E Susceptibilities.

- Review indicators of friendly COA.
- Use the products developed earlier in the C-SIGINT process to determine where threat capabilities and intentions are directed against susceptible MAGTF operations.
- Determine the probability of threat activity against MAGTF C-E operations.

Determine which Susceptibilities are Vulnerabilities.

- Designate as vulnerabilities those C-E susceptibilities targetable by a specific threat collector.
- List (and maintain separately) nontargetable indicators.
- Match indicators with threat systems and document specific event characteristics if known; for example, time and location of vulnerabilities.

Task 9—Rank Vulnerabilities

C-SIGINT analysts rank vulnerabilities by analyzing them in view of the indicators of friendly COAs and EEFIs. The ranking is based on criteria estimating the uniqueness, degree of susceptibility, and importance of the vulnerability. Analysts designate the vulnerability as critical, significant, or important to the success of the overall operation. The procedures for ranking vulnerabilities follow.

Establish Criteria for Measuring the Vulnerability. Develop a means for judging whether each identified vulnerability is critical, significant, or important to the success of the operation. These final ratings are attained by evaluating each vulnerability against criteria that address how critical they are to the success or failure of the operation. Uniqueness, importance, and susceptibility to threat are three criteria that measure vulnerability and criticality, and permit an accurate ranking of them. They are defined as follows:

- Uniqueness—the extent vulnerability can be readily associated with a COA.
- Importance—a measure of how critical vulnerability is to the success of the operation.
- Susceptibility to threat—a measure of the number and variety of threats placed against the indicator.

Compare Vulnerabilities to Criteria.

- Combine criteria and vulnerabilities in a matrix format shown in table C-3. For each vulnerability, conduct a review against the established criteria. Analysts have in their possession the commander's objectives, prioritized EEFI, ranking criteria, and can evaluate vulnerabilities using these data. Vulnerabilities are first rated according to each of the criteria. The horizontal axis of the matrix lists the criteria of uniqueness, importance, and susceptibility.
- List vulnerabilities on the vertical axis. The degree of satisfaction of a criterion is expressed numerically on a scale of 0 to 5 with 5 being the highest rating. If vulnerability is highly unique, pertaining to very specialized and infrequently exhibited indicators; it would be assigned a high rating. If vulnerability is exhibited in many COAs, in many operations, its uniqueness rating would be low (0 to 2).
- If a vulnerability is highly important, involving disclosure of a critical EEFI, its rating would be high. An EEFI lower on the commander's list of priorities would receive a lower rating. If vulnerability is highly susceptible, targeted by numerous threat systems of several types, its rating for susceptibility would be high.
- If a single threat system of limited capability is targeting the vulnerability, the rating would be low. The overall ratings are determined by adding the values of the three criteria and placing it under the overall number rating.

Table C-3. Vulnerability Matrix Format.

Vulnerability	EEFI	Criteria			Numerical Rating
		Uniqueness	Importance	Susceptibility	
Radios at main CP vulnerable to DF	4(a)2	5	5	4	14
Radios at main CP vulnerable to jamming	4(a)4	3	2	3	8

Criteria rating values	Overall rating values
0-2 = Low	0-4 = Unimportant
3 = Medium	5-8 = Important
4-5 = High	9-11 = Significant
	12-15 = Critical

Develop Ranking.

- Develop a prioritized ranking once an overall rating is established for each vulnerability. Vulnerabilities fall into the broader categories of critical, significant and important, based on the criticality level of criteria satisfied. Vulnerabilities receiving overall ratings between 5 and 8 are considered important; those between 9 and 11 are significant; and those falling between 12 and 15 would be critical.
- Enter the list of ranked vulnerabilities in the database. It is retained in hard copy for dissemination, and applied in the countermeasures options development in step three of the C-SIGINT process.

Task 10—Produce Output From Vulnerability Assessment.

The CI analyst presents the vulnerability assessment findings as a briefing or a report to the commander, G-3/S-3, unit security manager, and other key staff members.

Within a MAGTF, C-SIGINT analysis and production generally results from the integrated operations of the P&A cell, the radio battalion OCAC, and the supporting CI/HUMINT company CP.

Appendix D. Counterintelligence Plans, Reports, and other Formats

This page intentionally left blank

Section 1

Counterintelligence Estimate

Purpose. Provides a baseline of historical, threat related information for inclusion as Tab A of Appendix 3, CI to Annex B, Intelligence.

CLASSIFICATION

Counterintelligence Estimate

Copy no. __ of __ copies
ISSUING HEADQUARTERS
PLACE OF ISSUE
Date/Time Group
Message reference number

COUNTERINTELLIGENCE ESTIMATE (Number) (U)

(U) REFERENCES:

a. Unit SOP for intelligence and CI.

b. JTF, NTF, other components, theater, national intelligence and CI plans, orders, and tactics, techniques and procedures; and multinational agreements pertinent to intelligence operations.

c. Maps, charts, and other intelligence and CI products required for an understanding of this annex.

d. Documents and online data bases that provide intelligence required for planning.

e. Others as appropriate.

1. (U) Mission. (State the assigned task and its purpose.)

2. (U) Characteristics of the Area of Operations. (State conditions and other pertinent characteristics of the area that exist and may affect enemy intelligence, sabotage, subversive, and terrorist capabilities and operations. Assess the estimated effects. Also, assess their effects on friendly CI capabilities, operations, and measures. Reference appendix 8, Intelligence Estimate, to annex B, Intelligence, as appropriate.)

a. (U) Military Geography

(1) (U) Existing situation.

(2) (U) Estimated effects on enemy intelligence, sabotage, subversive and terrorist operations and capabilities.

(3) (U) Estimated effects on friendly CI operations, capabilities and measures.

Page number

CLASSIFICATION

b. (U) Weather

(1) (U) Existing situation.

(2) (U) Estimated effects on enemy intelligence, sabotage, subversive, and terrorist operations and capabilities.

(3) (U) Estimated effects on friendly CI operations, capabilities, and measures.

c. (U) Other Characteristics. (Additional pertinent characteristics are considered in separate subparagraphs: sociological, political, economic, psychological, and other factors. Other factors include but are not limited to telecommunications material, transportation, manpower, hydrography, science, and technology. These are analyzed under the same headings as used for military geography and weather.)

3. (U) Intelligence, Sabotage, Subversive, and Terrorist Situation. (Discusses enemy intelligence, sabotage, subversive, and terrorist activities as to current situations and recent/significant activities. Include known factors on enemy intelligence, sabotage, subversive, and terrorist organizations. Fact sheets containing pertinent information on each organization may be attached to the estimate or annexes.)

a. (U) Location and disposition.

b. (U) Composition.

c. (U) Strength. (Including local available strength, availability of replacements, efficiency of enemy intelligence, sabotage, subversive, and terrorist organizations.)

d. (U) Recent and present significant intelligence, sabotage, and subversive activities/movements. (Including enemy knowledge of our intelligence and CI efforts).

e. (U) Operational, tactical, and technical capabilities and equipment.

f. (U) Peculiarities and weaknesses.

g. (U) Other factors as appropriate.

4. (U) Intelligence, Sabotage, Subversive, and Terrorist Capabilities and Analysis. (List separately each indicated enemy capability that can affect the accomplishment of the assigned mission. Each enemy capability should contain information on what the enemy can do, where they can do it, when they can start it and get it done, and what strength they can devote to the task. Analyze each capability in light of the assigned mission, considering all applicable factors from paragraph 2, and attempt to determine and give reasons for the estimated probability of adoption by the enemy. Examine the enemy's capabilities by discussing the factors that favor or militate against its adoption by the enemy. Analysis of each capability should include a discussion of enemy strengths and vulnerabilities associated with that capability and a discussion of any indications that point to possible adoption of the capability. Finally, state the estimated

Page number

effect the enemy's adoption of each capability will have on the accomplishment of the friendly mission.)

a. (U) Capabilities

(1) (U) Intelligence. (Include known/estimated enemy methods.)

(2) (U) Sabotage. (Include possible agent/guerilla capabilities for military, political, and economic sabotage.)

(3) (U) Subversion. (Include all types, such as propaganda, sedition, treason, disaffection, and threatened terrorists activities affecting our troops, allies, and local civilians, and assistance in the escape and evasion of hostile civilians.)

(4) (U) Terrorist. (Include capabilities of terrorist personalities and organizations in AO.)

b. (U) Analysis and discussion of enemy capabilities for intelligence, sabotage, subversive, and terrorism as a basis to judge the probability of their adoption.

5. (U) Conclusions and Vulnerabilities. (Conclusions resulting from discussion in paragraph 4. Relate to current all-source intelligence estimates of the enemy's centers of gravity, critical and other vulnerabilities and estimated exploitability of these by friendly forces, enemy courses of action beginning with the most probable and continuing down the list in the estimated order of probability, and the estimated effects adoption of each capability would have on the friendly mission.)

a. (U) Probability of enemy adoption of intelligence, sabotage, subversive, and terrorist programs or procedures based on capabilities.

b. (U) Effects of enemy capabilities on friendly course of action.

c. (U) Effectiveness of our own CI measures and additional requirements or emphasis needed.

Name
Rank and Service
Title

EXHIBITS:
(As appropriate)

DISTRIBUTION:

This page intentionally left blank.

Section 2

Counterintelligence Reduction Plan

Purpose. A visual working tool for managing the CI targeting triad—PO&I—and unit assignments.

CLASSIFICATION

Counterintelligence Reduction Plan

PERSONALITY/INSTALLATION: Operation Bold Lighting MAP REF: Name, Series

Target Number	Target	Location/ Description	PRI	CI Team Assign	Special Instructions	Interested Units/Sect
1	Broad-casting Station	Grid coordinates 3 km NW of city on Victory Road	1	1	Locate/take into custody station state security officer and all propaganda file material	G-5/PAO
2	Government Control Center	Grid coordinates largest building in city center, gable roof	3	1	Ensure file information protected for further analysis	G-5
3	Military Intelligence Headquarters	Grid coordinates located on liberation military compound E of city	1	2	Locate/search CI and agent operations section	MAGTF G-2
4	Smith, John Q Intelligence Cadre	Gird coordinates military intelligence headquarters (above). Home address: 134 8th St, Apt 3B	2	2	Potential defector handle accordingly	1st MARDIV
5	Infiltration Training Facility	Grid coordinates located W of city on seaward peninsula	3	2	Secure/search for file information on personalities and operations. Coordinate with Task Force N2	MAGTF G-2

Page number

CLASSIFICATION

CLASSIFICATION

Target Number	Target	Location/ Description	PRI	CI Team Assign	Special Instructions	Interested Units/Sect
6	Political Prison	Grid coordinates triangular shaped compound enclosed by 20-foot block wall bordered by Liberation Ave, 9th St and bay	2	1	Personalities of CI interest on separate listing. Provide list of recovered personalities to HQ via most expedient means	G-5/G-3/G-4
7	National Intelligence Field Office	Gird coordinates located in concrete block building on corner of 5th St and Liberation Ave	1	2	Immediately evaluate all documents/ equipment	G-2

Page number

CLASSIFICATION

Section 3

Counterintelligence Salute Report Format

Purpose. A quick response report to get information into the all-source intelligence data base.

CLASSIFICATION

Counterintelligence Salute Report Format

Reporting Unit: (Text Field)

Record Creator: (Text Field)

Report Number: (Text Field)

References: (Text Area)

Requirement Reference: (Text Field)

Size (of Enemy Unit): (Text Field)

Activity Type: (Text Field w/Picklist)

Activity Status: (Text Field w/Picklist)

Activity Location: (Text Field)

Map Coordinates:
(Text Fields for—
- Latitude
- Longitude
- Map Grid Reference
- UTM)

Activity Direction: (Text Field)

Unit: (Text Field)

Date Event Began/To Begin: (Text Field)

Time Event Began/To Begin: (Text Field)

Date Event Ended/Expected to End (Text Field)

Time Event Ended/Expected to End (Text Field)

Equipment: (Text Area)

SRC #: (Text Field)

SRC Description: (Text Area)

SRC Reliability: (Text Field w/Picklist)

Comments: (Text Area)

Map Data: (Text Field)

Page number

CLASSIFICATION

This page intentionally left blank.

Section 4

Counterintelligence Information Report

<u>Purpose</u>. A standard report used to report tactical CI information.

CLASSIFICATION

Counterintelligence Information Report

Record ID: ....
Point of Contact:..
Classification:..
Abstract: ....
Discretionary Access Control:......................................
Caveats: ....
Release To: ....
Record Type: ....
Record Status:...
Date Created (yyyymmdd): ....
Date Modified (yyyymmdd):.......................................
Community of Interest:...
Source Record: ....
Requirement Reference: ....
Requirement: ....
Title (Text):...
Report Number: ....
Report Date (yyyymmdd):..
To:...
Target:..
Individual Source: ....
Reliability of the Source:..
Source ID Number:...
Information Reliability: ....
Information Date (yyyymmdd): ....
Collection Date (yyyymmdd):.....................................

Page number

CLASSIFICATION

CLASSIFICATION

Location: ....

Report (Text): ....

Comments (Text): ....

Page number

CLASSIFICATION

Section 5

Counterintelligence Force Protection
Source Operations Concept Proposal

Purpose. Serves as the planning, approval, and execution vehicles for MAGTF CI Force Protection Source Operations.

CLASSIFICATION

Counterintelligence Force Protection
Source Operations Concept Proposal

From: (Originator) (Text Field-Picklist?)

To: (Next Higher Echelon) (Text Field-Picklist?)

Info: (Addees to be informed) (Text Area-Picklist?)

Project Number: (Text Field)

Name: (Text Field)

Originating Hqs: (Text Field)

Implementing Element: (Text Field)

Collection Requirements: (Text Area)

References: (Text Area)

Date Submitted: (Text Field)

Date Approved: (Text Field)

Approval Authority: (Text Field)

Operation Type: (Text Field w/Picklist)

Target Focus: (Text Area)

Target Personnel: (Text Area)

Target Country: (Text Area)

Organizations: (Text Area)

Base of Operations: (Text Field)

Communications Methods: (Text Area)

Risks: (Text Area)

Technical Support: (Text Area)

Finances: (Text Area)

Page number

CLASSIFICATION

CLASSIFICATION

Coordination: (Text Area)

Administration and Management: (Text Area)

Additional Support Requirements: (Text Area)

Point of Contact: (Text Area)

Page number

CLASSIFICATION

Section 6

Counterintelligence Source Lead Development Report

Purpose. For local planning and development of CI sources.

CLASSIFICATION

Counterintelligence Source Lead Development Report

Date: (Text Field) (Mandatory)

Subject: (Text Field) (Mandatory)

Report No: (Text Field)

Project No: (Text Field)

References: (Text Field)

Record Creator: (Text Field) (Mandatory)

Origin: (Text Field)

Source of Lead: (Text Field) (Mandatory)

Proposed Use of Lead: (Text Area)

Lead Screening Process: (Text Area)

Placement and Access: (Text Area)

Circumstances for Meeting with Source: (Text Area)

Security Issues: (Text Area)

Personnel Information: (Text Area)

Lead Status: (Text Area)

Nationality: (Text Area)

Citizenship: (Text Field w/PL)

Personality and Character Traits: (Text Field w/PL)

Motivation: (Text Field)

Character: (Text Field)

Personality: (Text Area)

Trait Exploitation: (Text Area)

Biographical Data: (Link to the INDIVIDUAL record. When printing, whole INDIVIDUAL record needs to print out.)

Summary of Family/Personal History: (Text Area)

(Consider autopopulate from INDIVIDUAL record.)

Investigative Checks:

Type (Text Field w/PL) Status (Text Field w/PL) Date (Text Filed w/PL)

Page number

CLASSIFICATION

Coordination Required: (Text Field w/PL) (Multiple Choice)

Assessment of Operational Potential:

Type of Source: (Text Field w/PL)

Placement: (Text Area)

Access: (Text Area)

Cover (For Status & Action): (Text Area)

Qualifications: (Text Area)

Personal: (Text Area)

Strengths and Weaknesses: (Text Area)

Risk:

To C/O & Collection Element: (Text Area)

To Source: (Text Area)

Approach Plan: (Text Area)

ICF: (Text Area)

Comments: (Text Area)

Attachments: (Standard Repeating Group)

Section 7

Counterintelligence Screening Report

Purpose. Used to report information obtained during CI screening operations.

CLASSIFICATION

Counterintelligence Screening Report

Reporting Unit: (Text Field)

Screener: (Text Field)

Report Date: (Text Field)

Report Time: (Text Field)

Capturing Unit: (Text Field)

Requirement Reference: (Text Field)

Status: (Text Field w/PL)

PL—Military
Paramilitary
Civilian
Other

Name: (Text Field)

Alternate Name(s): (Repeating Group)

Personal ID No: (Text Field)

EPW ID No: (Text Field)

Date of Birth: (Text Field)

Sex: (Text Field w/PL)

Marital Status: (Text Field)

Language Competence: (Use text and field input from DCIIS Individual form)

Language Used: (Text Field)

Education: (Use text and field input from DCIIS Individual form)

Employment: (Use text and field input from DCIIS Individual form)

Military Service: (Use text and field input from DCIIS Individual form)

Date Captured: (Text Field)

Time Captured: (Text Field)

Place Captured: (Text Field)

Circumstances of Capture: (Text Area)

Documents at Capture: (Text Area)

Page number

CLASSIFICATION

CLASSIFICATION

Equipment Captured: (Text Area)

Source's Physical Condition: (Text Field w/PL)

Remarks: (Text Area)

Source's Mental State: (Text Field w/PL)

Source's Intelligence Level: (Text Field w/PL)

Specific Knowledgeability: (Text Area)

Source's Cooperation: (Text Field w/PL)

EPW Category: (Text Field w/PL)

CI Interest: (Text Field w/PL)

Source's Current Location: (Text Field)

Approach Plan: (Text Field w/PL)

Comments: (Text Area)

Page number

CLASSIFICATION

Section 8

Counterintelligence Tactical Interrogation Report

CLASSIFICATION

Counterintelligence Tactical Interrogation Report

Reporting Unit: (Text Field)

Report No: (Text Field) Record Creator: (Text Field)

Report Date: (Text Field) Interpreter: (Text Field)

Report Time: (Text Field) Language Used: (Text Field w/PL)

Capturing Unit: (Text Field)

Requirement Reference: (Text Field)

Map Data: (Text Area)

Source No: (Text Area)

Source Status: (Text Field w/PL)

Name: (Text Field)

Alternate Name(s): (Repeating Group)

Personal ID No: (Text Field)

EPW ID No: (Text Field)

Place of Birth: (Text Field)

Date of Birth: (Text Field)

Nationality: (Text Field)

Sex: (Text Field w/PL)

Marital Status: (Text Field)

Language Competence: (Use text and field input form DCIIS Individual from)

Language Used: (Text Field)

Education: (Use text and field input form DCIIS Individual from)

Employment: (Use text and field input form DCIIS Individual from)

Military Service: (Use text and field input form DCIIS Individual from)

Date Captured: (Text Field)

Time Captured: (Text Field)

Place Captured: (Text Field)

Circumstances of Capture: (Text Area)

Documents at Capture: (Text Area)

Equipment Captured: (Text Area)

Page number

CLASSIFICATION

CLASSIFICATION

Source's Physical Condition: (Text Field w/PL))

Source's Mental State: (Text Field w/PL)

Source's Intelligence Level: (Text Field w/PL)

Specific Knowledgeability: (Text Area)

Source's Cooperation: (Text Field w/PL)

EPW Category: (Text Field w/PL)

CI/HUMINT Interest: (Text Field w/PL)

Source's Current Location: (Text Field)

Source's Reliability: (Text Field)

Source's Production: (Text Field)

Approach Plan: (Text Field w/PL)

Comments: (Text Area)

Page number

CLASSIFICATION

Section 9

Intelligence Information Report

Purpose. Standard report used to report unevaluated, unanalyzed intelligence information.

CLASSIFICATION

Intelligence Information Report

From: (Text Field)

To: (Text Field)

Info: (Text Field)

Serial: (Text Field)

Country: (Text Field)

//IPSP: (Text Field)

Subj: (Text Field)

WARNING: (U) THIS IS AN INFORMATION REPORT, NOT FINALLY EVALUATED INTELLIGENCE. REPORT CLASSIFIED (Autopopulate with classification)

DEPARTMENT OF DEFENSE

DOI: (Text Field)

REQS: (Text Field) (Association Mechanism)

SOURCE: (Text Field)

SUMMARY: (Text Field)

TEXT: (Text Field)

COMMENTS: (Text Area)

(FIELD COMMENT) (Text Area)

PROJ: (Text Field)

INSTR: US NO: (Text Field)

PREP: (Text Field)

ENCL: (Text Field) (Repeating Group)

ACQ: (Text Field)

DISSEM: FIELD—(Text Field)

WARNING: REPORT CLASSIFIED (Text Field) (Autopopulate)

DRV FROM—(Text Field)

DECL: (Text Field)

Page number

CLASSIFICATION

This page intentionally left blank.

Section 10

Intelligence Information Report—Biographical

Purpose. Standard report used to report unevaluated, unanalyzed biographical intelligence information.

CLASSIFICATION

Intelligence Information Report—Biographical

From: (Text Field)

To: (Text Field)

Info: (Text Field)

Serial: (Text Field)

Country: (Text Field)

//IPSP: (Text Field)

Subj: (Text Field)

WARNING: (U) THIS IS AN INFORMATION REPORT, NOT FINALLY EVALUATED INTELLIGENCE. REPORT CLASSIFIED (Autopopulate with classification)

DEPARTMENT OF DEFENSE

DOI: (Text Field)

REQS: (Text Area) (Association Mechanism)

SOURCE: (Text Area)

SUMMARY: (Text Area)

TEXT:

1. Name of Country (Text Field w/PL)

2. Date of Information (Text Field)

3. Date of Report (Text Field)

4A. Full Name (Text Field)

4B. Name(s) By Which Individual Prefers To Be Addressed

4B(1). In Official Correspondence (Text Field)

4B(2). Orally at Official Gatherings (Text Field)

4C. Full Name in Native Alphabet (Text Field) (In standard telegraphic code or other transcription code)

4D. Variants, Aliases or Nicknames (INDIVIDUAL Record repeating group autopopulate)

Rank

Page number

CLASSIFICATION

5A. English Language (Text Field)

5B. Native (Text Field)

Date of Rank (Text Field)

Position/Billet (Text Field)

7A. Present Position (Text Field)

7B. Military Address (Text Field)

7C. Date Assumed Position (Text Field)

7D. Scheduled Date of Departure (Text Field)

7E. Name of Predecessor

7E1. Predecessor's Name (Text Field)

7E2. Predecessor's Branch of Service (Text Field)

7E3. Date Predecessor Assigned (Text Field)

7E4. Duration of Predecessor's Assignment (Text Field)

Branch of Armed Service (Text Field)

Specialty/Other Organizations

Date of Birth (Text Field)

Place of Birth (Text Field)

Sex (Text Field)

Home Address (Text Fields) (autopopulate)

Telephone Number

14A. Home (Text Fields)

14B. Work (Text Fields)

Marital Status (Text Fields)

Citizenship (Text Fields)

COMMENTS:

(FIELD COMMENT) (Text Area)

PROJ: (Text Field)

INSTR: US NO (Text Field)

PREP: (Text Field)

ENCL: (Text Field) (Repeating Group)

ACQ: (Text Field)

DISSEM: FIELD—(Text Field)

WARNING: REPORT CLASSIFIED (Text Field) (Autopopulate)

DRV FROM—(Text Field)

DECL: (Text Field)

Page number

CLASSIFICATION

Section 11

Counterintelligence Inspection/Evaluation Report

CLASSIFICATION

Counterintelligence Inspection/Evaluation Report

(Normally hard copy report—not templated in DCIIS)

Reporting Unit:
Dissemination:
Report Date:
Report Time:
Reference:
Enclosure:

Synopsis: (Summary of the report)

1. (U) Predication. (What initiated the inspection/evaluation).

2. (U) Purpose. (What the inspection/evaluation was to determine. State any limitations that were placed on the activity.)

3. (U) Background. (Information on previous inspections/evaluations or surveys on the same area. Information on level and amount of classified material maintained. Identity of person(s) conducting the activity.)

4. (U) Results. (Detailed information obtained during the inspection/evaluation. Describe security measures in effect, whether the measures required by appropriate references were adequate, and any identified security weaknesses/deficiencies.)

5. (U) Recommendations. (List recommendations to correct security weaknesses or deficiencies as they appear in paragraph 4 above, reference the paragraph for clarity.)

(Signature on line above typed name)
REPORTED BY. (Typed name of evaluator)

(Signature on line above typed name)
APPROVED BY. (Typed name and title of approving authority)

Page number

CLASSIFICATION

This page intentionally left blank.

Section 12

Counterintelligence Survey/Vulnerability Assessment

CLASSIFICATION

Counterintelligence Survey/Vulnerability Assessment

(Normally hard copy report—not templated in DCIIS)

Reporting Unit:
Dissemination:
Report Date:
Report Time:
Reference:
Enclosure:

SYNOPSIS: (Summary of the report)

1. (U) Predication. (How the survey was initiated.)

2. (U) Purpose. (What the survey was to determine. State any limitations on the survey.)

3. (U) Background:

   a. (U) Person(s) conducting the survey.

   b. (U) Previous surveys.

   c. (U) Mission.

   d. (U) Inherent hazards of the area.

   e. (U) Degree of security required (Maximum, medium, or minimum based on the following factors):

      (1) (U) Mission

      (2) (U) Cost of Replacement

      (3) (U) Location

      (4) (U) Number of like installations

      (5) (U) Classified Material

      (6) (U) Importance

Page number

CLASSIFICATION

4. (U) Results:

a. (U) Security of information

b. (U) Security of personnel

c. (U) Physical security

5. (U) Recommendations. (List recommendations to correct security hazards as they appear in paragraph 4 under the subparagraph heading, reference the paragraph for clarity.)

a. (U) Security of information.

b. (U) Security of personnel.

c. (U) Physical security.

(Signature on line above typed name)
REPORTED BY. (Typed name of evaluator)

(Signature on line above typed name)
APPROVED BY. (Typed name and title of approving authority)

Page number

Section 13

Counterintelligence Survey/Vulnerability Assessment Checklist

CLASSIFICATION

Counterintelligence Survey/Vulnerability Assessment Checklist

Background:

CI surveys/vulnerability assessments are conducted during peacetime as well as times of hostilities, both in and outside the continental U.S.

Manning of CI survey/vulnerability assessment teams should be task-organized to meet the needs/requirements of the survey; i.e., CI officers/specialists, physical security specialists from the provost marshal, communications specialists, data processing security specialists, etc.

Name of Installation
Location of Installation
Type of Installation

1. Functions, Purpose or Activities at Installation

   a. What troops, units, and command elements are stationed there or use or control the installation?

   b. What military activities (conventional, unconventional, or special) take place?

   c. What material is produced, processed, tested, or stored?

   d. What is the military importance?

2. Critical Rating of the Installation

   a. How important to national security or Marine Corps forces are the activities that take place at the installation?

   b. What activity on the installation should be veiled in secrecy? Why?

   c. What information about the installation would be of interest to hostile intelligence? Why?

   d. Is this the only location where the activities taking place can be conducted?

   e. Are there substitute places available that are suitable and practical?

Page number

CLASSIFICATION

f. Is there a key facility/organization aboard the installation?

g. Is there any sensitive or critical material or equipment stored, tested, or developed aboard the installation?

h. Is the installation a likely target for espionage?

3. Names of Principal Officers of the Installation/Organization

4. Names of Persons Directly Responsible for the Security of the Installation

5. Physical Location and Description of the Installation

a. This is the physical description of the general area surrounding the installation, paying particular attention to road networks, rail facilities, air facilities, transportation, terrain, etc.

b. Include a general physical description of the entire installation, to be accompanied wherever possible by a map, sketch or aerial photograph, and the following information:

(1) Area and perimeter.

(2) Numbers, types, and locations of buildings, and relationships among the various buildings.

(3) Roads, paths, railroad sidings, canals, rivers, etc., on the premises of the installation.

(4) Wharves, docks, loading platforms, etc., on the premises.

(5) Any other distinctive structures or features.

c. Note any particularly vulnerable or sensitive points on the installation, and the reasons for their vulnerability or sensitivity. Pay particular attention to the following:

(1) Command element/headquarters buildings.

(2) Operations/crisis action facilities.

(3) Repair shops (armor, vehicle, aircraft, weapons, and communications).

(4) Power plants.

(5) Transformer stations.

(6) Warehouses.

(7) Communications systems/facilities.

(8) Fuel storage.

(9) Water tanks, reservoirs, supply systems.

(10) Motor pools.

(11) Ammunition dumps.

(12) Aircraft.

(13) Firefighting equipment.

(14) Military police/reaction force location and reliability.

(15) Special training/testing sites.

6. Perimeter Security

a. Description of the perimeter and physical barriers.

(1) What type of fence or other physical barrier around the installation affords perimeter security?

(2) Describe the construction material of the fence/barrier.

(3) How high is the fence/barrier?

(4) Is the fence/barrier easily surmountable?

(5) Is the top protected by barbed wire outriggers?

(6) Are there any cuts, breaks, tears, holes, or gaps in the fence/barrier or any holes under it?

(7) Are there any tunnels near or under the fence/barrier?

(8) Are vehicles parked near or against the fence/barrier?

(9) Are piles of scrap, refuse or lumber kept near the fence/barrier?

(10) Is the fence/barrier patrolled and checked daily for cuts, breaks, holes, gaps, tunnels, or evidence of tampering?

(11) Where are pedestrian and vehicle gates located?

(12) Are unguarded gates firmly and securely locked?

(13) Are the gates constructed in a manner where identity and credential checks of persons or vehicles entering or exiting are accomplished, particularly during rush hours?

(14) During what hours is each gate open?

(15) Are there any railroad rights of way, sewers, tree lines, or other weak points on the perimeter?

(16) Are these weak points guarded, patrolled, or secured in any fashion?

(17) Is high intensity lighting used to light up the perimeter during hours of darkness?

(18) Where are the lights located?

(19) Are there dead spots between lighted areas?

(20) Is there backup emergency power for the lighting?

(21) Does the lighting inhibit/hamper security force observation?

7. <u>Perimeter Security Force</u>

a. Description of the organization of the security force.

(1) What is the strength of perimeter security forces?

(2) What are the number and location of guard posts?

(3) What is the length of perimeter covered by each post?

(4) What is the length of watch of each post?

(5) Is there a reserve backup or security force?

(6) What weapons do the guards carry?

(7) What is the level of training for each member of the security force?

(8) What instructions are given to security forces regarding identity checks/challenges?

(9) Are there vehicle checks?

(10) Are there any watchtowers to facilitate observation of the perimeter?

(11) What are the height and location of each watchtower?

(12) Have roving patrols been utilized to patrol the perimeter? What are their number, strength, frequency, routes, and activity?

(13) What is the efficiency and manner of performance of perimeter guards and patrols?

b. Security weaknesses and recommendations.

(1) What specific weaknesses pertaining to both physical barriers and the perimeter security force were noted during the survey?

(2) What specific and reasonable recommendations may be made to improve perimeter security?

8. Security of Buildings and Structures

a. Nature and purpose of building.

(1) Where is the location of the building?

(2) What activities take place in the building?

(3) What material/information is developed or stored?

(4) What machinery or equipment is in the building?

(5) Is the building a vulnerable point? Why?

b. Description of building—exterior, interior, and immediate surroundings.

(1) Describe the design and construction of the building.

(2) How many stories? Height?

(3) Does the building have a basement?

(4) What percentage is wood?

(5) What percentage is concrete?

(6) What other materials are used in the exterior?

(7) Describe walls.

(8) Describe floors.

(9) Describe ceilings.

(10) Describe roof.

(11) Is the building safely designed and constructed?

(12) Is the building properly maintained and constructed?

(13) Check locations of doors, windows, sewers, sidewalks, elevators, stairs, fire escapes, skylights, crawl spaces/false ceilings, and any other possible means of exit or entry.

(14) Are these entrances properly locked, or otherwise safeguarded against unauthorized entry?

(15) Are windows and skylights screened, grilled, or barred?

(16) Can unauthorized or surreptitious entry be effected in any manner?

(17) Are exit and entry facilities adequate to meet an emergency situation?

(18) Are all keys to building controlled?

(19) Where is the key control maintained?

(20) Who maintains the key control?

(21) How rigorously is it kept?

(22) Who is authorized to receipt for the keys?

(23) Are any measures taken to restrict entry into the building; i.e., pass, badges, access rosters, etc.

(24) Are controlled access methods enforced?

(25) If the building is determined to be sensitive, high threat priority, or vulnerable, has it been declared as restricted, and is the area surrounding it so designated?

(26) Are daily security checks conducted at the end of each working day in areas where classified material is stored? Are all security containers checked?

c. Guard and patrol system around the building.

(1) What are the duties of guards and patrols?

(2) Are high intensity lights used to light up the exterior and the area surrounding the buildings during hours of darkness?

Page number

CLASSIFICATION

(3) Is there a reactionary security force?

(4) What is the response time? Has it been tested?

(5) What is the size of the guard force? Reactionary force?

(6) What are the means of activating the reactionary force? Are there backup systems?

d. Security of electrical equipment.

(1) Is there auxiliary lighting?

(2) Are circuit breakers properly protected?

(3) Are telephone junction boards protected?

e. List the frequency of periodic checks made throughout the building to detect the following:

(1) Holes, cracks, crevices that might conceal explosives, incendiary devices, or audio/visual monitoring devices. Are such repaired?

(2) Tampered wiring, or broken or electrical connections and wires.

(3) The presence of suspicious packages or bundles.

(4) Any dangerous practices, including safety, electrical, or fire hazards that may result from negligence or deliberate attempts at sabotage,

f. Security weaknesses and recommendations.

(1) What specific weaknesses pertaining to the security of interiors and exteriors were noted during the survey?

(2) What specific reasonable recommendations may be made to improve the security of the buildings?

9. Security of Docks, Wharves, and Platforms

a. Description of the location, nature, and purpose of each dock, wharf, or platform.

(1) What administrative supervision of the docks, wharves, and loading platforms is exercised? By whom?

(2) What type of security force provides protection for each?

(3) What measures are taken to prevent loitering in the vicinity of each?

(4) What measures are taken to prevent unauthorized observation of loading and unloading?

(5) What protection is afforded mechanical sabotage, arson, explosion, or dangerous practices?

(6) Are same precautionary measures taken as outlined for access to building interiors and exteriors?

b. Traffic conditions.

(1) Are inspections of deliveries made to guard against sabotage devices; i.e., explosives, caustic chemicals, etc.?

(2) What precautions are taken to conceal the loading and unloading of personnel or material if such handling requires secrecy?

(3) Are delivery trucks, railroad cars, and privately owned vehicles checked for possible sabotage devices?

(4) How much is the movement of drivers and helpers aboard the installation controlled?

(5) In the case of movement of personnel, equipment, and material, are identifying markings removed in an effort to assist in operations security?

c. Security weaknesses and recommendations.

(1) What specific weaknesses pertaining to the security of docks, wharves, and loading platforms were noted during the survey?

(2) What specific and reasonable recommendations may be made to improve the security of docks, wharves, and loading platforms?

10. <u>Motor Pools, Dismount Points, and Parking Areas</u>

a. Security measures at each facility.

(1) Are motor pools, dismount points, and parking areas adequately guarded?

(2) Are vehicles properly checked and accessible only to authorized personnel?

(3) What system of checking vehicles is used?

(4) What measures are taken to safeguard fuels, lubricants, tools, and equipment against sabotage, theft, fire, and explosion?

(5) Are frequent checks made of all vehicles for possible mechanical sabotage?

(6) Are drivers and mechanics instructed as to the proper checks to be made to guard against or detect sabotage?

(7) What provisions are made to prohibit privately owned vehicle parking in motor pools, dismount points/parking areas?

(8) Are fuels and lubricants frequently tested for possible contamination?

(9) Are parking/staging areas restricted or supervised in any way?

(10) Are parking arrangements consistent with security against sabotage, terrorist, or other hazards?

(11) What provisions are made for visitors parking?

(12) Do parking arrangements/facilities impede efficient traffic flow through and near the compound?

(13) Would parking arrangements interfere with firefighting or other necessary emergency vehicles if there were an emergency?

b. Security weaknesses and recommendations.

(1) What specific weaknesses about the security of motor pools and parking lots were noted during this survey?

(2) What specific and reasonable recommendations may be made to improve the security of the motor pool and parking lots.

11. <u>Power Facilities and Supply</u>

a. Description of supply, facilities, and security measures.

(1) What type of power is used by the installation?

(2) What is the peak load of electric power?

(3) What percentage of the electric power is generated on the installation?

(4) What is the installation's electric generating capacity?

(5) What percentage of electric power is purchased from outside sources?

(6) Are all current sources ample to provide a reserve beyond full load demands?

(7) From whom is the electric power purchased?

(8) Is an alternate or auxiliary electric power system available for emergency use?

(9) Can the auxiliary electric power system be used immediately?

(10) How many and what kind of power substations/transformers are on the installation?

(11) Are control panels, pressure valves, gas facilities, and control valves in good working order? How frequently are they checked? Is adequate fire protective equipment available and nearby?

(12) Are power substations/transformers adequately safeguarded against trespassers and saboteurs?

(13) Are generators properly maintained and checked with particular emphasis on oil levels and temperatures?

(14) Are combustible materials removed from their vicinity?

b. Miscellaneous features:

(1) Are replacement units for generators and motors available in safe storage?

(2) Do transformers have sufficient capacity? Are they safely located and well-protected by physical barriers and guards?

(3) Are oil-filled transformers located in noncombustible well-drained buildings or outside?

(4) Are frequent inspections made of the oil, contact, and control apparatus of circuit breakers and transformers?

(5) What is the system of power lines in use?

(6) What is the number of independent power feeds?

(7) Is the pole line or underground line safe, reliable, and frequently checked?

(8) Are all power lines protected by lightning arresters?

(9) Are power distribution lines properly installed and supported?

(10) Are electric circuits overloaded at any time?

(11) Are current national or civil electric codes followed?

(12) Is there a single or multiple main switch(s) for emergencies?

c. Security weaknesses and recommendations.

(1) What specific weakness about the security of power facilities and supply were noted during the survey?

(2) What specific and reasonable recommendations may be made to improve the security of power facilities and supply?

12. Fire Fighting Equipment and Facilities

a. Describe the amount and condition of equipment and facilities.

(1) What fire fighting and first aid equipment are available on the installation?

(2) What types of fire extinguishers are available; i.e., foam, dry chemical, halon, water, carbon dioxide, and carbon tetrachloride? Are they at locations where such types may be needed?

(3) Are all extinguishers and other equipment in working order and frequently tested and inspected?

(4) Are fire extinguishers sealed to prevent tampering?

(5) Do competent personnel make inspections of fire equipment and record results recorded?

(6) Are both first aid and firefighting equipment painted to be conspicuous? Are they within reach of all personnel, unobstructed, and of reasonable size and weight to permit ease of handling by all personnel?

(7) Is first aid equipment available? Does it include ample amounts of materials needed?

(8) Are first aid supplies checked periodically and safeguarded?

(9) What type of fire alarm system(s) is/are installed?

(10) Are there sufficient numbers of alarms and sensors in the system?

(11) Is the fire alarm system frequently inspected and tested?

(12) Are vulnerable and/or important facilities equipped with sprinkler systems?

(13) What type of sprinkler system(s) is/are used? Are they fed by public mains, tanks, private reservoir, or pumps?

(14) How often and thoroughly is the sprinkler system inspected?

(15) Where are the main control valves of the system located?

(16) Are fire hydrants near vulnerable or important facilities?

(17) Are hydrants in working order? How often are they inspected and tested?

(18) Is the water pressure sufficient so that streams of water will reach and extinguish flames in all sections of the installation?

(19) Is there a secondary source of water supply available?

(20) Does the installation have its own fire department? A brigade? What equipment does it have? Are the personnel well trained?

(21) Have arrangements been made with public fire departments to furnish equipment and personnel to augment the installation department/brigade?

(22) Is the nearest public fire department paid or is it a volunteer unit?

(23) Has a program of fire drills been inaugurated? Are such drills conducted in an efficient and earnest manner?

(24) Has a fire prevention program been inaugurated?

(25) What plans have been made for the action of all personnel if there is a fire?

b. Security weaknesses and recommendations.

(1) What specific security weaknesses about the firefighting equipment and facilities were noted during the survey?

(2) What specific and reasonable recommendations may be made to improve the security of firefighting equipment and facilities?

13. Water Supply

a. Description of water supply and security measures taken to safeguard it.

(1) What sources of water supply are used by the installation?

(2) Are sources of water reasonably safe, adequately guarded, and protected by physical security?

(3) If a public supply is used, what is the diameter of the main line?

(4) What is the water pressure? Is it adequate for normal use as well as for emergencies?

(5) If a private reservoir or tank is used, what is its capacity, level, pressure, and condition?

(6) Is it adequate for the installation's needs?

(7) What type of pumps are used in the water system (underwater, suction, centrifugal, electric, etc.)?

(8) Are water pumping stations adequately protected, frequently inspected, and tested?

(9) Are all valves secured properly?

(10) Is a supplementary water system available? Where? Is it secure?

(11) How often is water tested for purification? By whom? Is the water treated? By whom? By what chemicals?

(12) Are taps/sources of unpotable water adequately marked?

(13) Is the sewage system adequate for the installation?

(14) Are sewer mains, control, pumps, and disposal systems adequate?

(15) Is there a possibility of water or food contamination from the sewage system?

(16) Has there been any epidemic outbreak at the installation traceable to waste disposal?

b. Security weaknesses and recommendations.

(1) What specific and reasonable recommendations may be made to improve the security of the water supply?

(2) What specific weaknesses about the water supply were noted during the survey?

14. Food Supply

a. Description of security measures.

(1) From what sources does the installation receive food and allied supplies? Can these sources be considered reliable?

(2) If food supplies are purchased from merchants and farmers in the local vicinity, have they been checked and tested for cleanliness?

(3) Have caterers and companies or individuals who operate food, candy, soft drink, or other concessions on or near the installation been checked? Have their products been thoroughly tested?

(4) Have local food handlers been checked for health, cleanliness, and loyalty?

(5) Is entry to kitchens and food storerooms restricted to authorized personnel? How are such restrictions enforced?

(6) Are pantries and refrigerators locked when not in use?

(7) Are kitchens and storerooms in sanitary condition?

(8) Is there any evidence of unsanitary conditions?

(9) Are frequent checks made of foods, drinks, etc., to prevent or detect toxicological or bacteriological sabotage?

(10) Has there been any epidemic or excess absenteeism traceable to food or water supplies of the installation?

b. Security weaknesses and recommendations.

(1) What specific weaknesses about the security of the food supply were noted during the survey?

(2) What specific and reasonable recommendations may be made to improve the security of the food supply?

15. Communications Facilities

a. General service and special communications message centers.

(1) Description.

(2) Where is the message center located?

(3) Is the message center adequately protected by physical barriers and guards?

(4) Is someone on duty at the message center at all times?

(5) Who handles the mail at the message center? Have mail handlers been subject to background and local records checks?

(6) Are all encryption (hardware and software) devices properly safeguarded and properly destroyed when obsolete?

(7) Are logs kept of authorized couriers and message traffic distribution?

(8) Are unauthorized personnel excluded from the message center?

(9) Are classified messages handled in accordance with OPNAVINST 5510.1

(10) Through what channels do classified messages pass?

(11) Have messengers, couriers, and operators been checked? Do they have appropriate security access(es)?

b. Security weaknesses and recommendations.

(1) What specific weaknesses about the security of the communications systems were noted during the survey?

(2) What specific and reasonable recommendations may be made to improve the security of the communications system?

16. Wire and Wireless Communications Equipment

a. Description.

(1) What means of wire and wireless communications are used throughout the installation?

(2) Where are the central points of such communications networks located?

(3) Are switchboards adequately guarded?

(4) Have operators been checked and cleared?

(5) Is auxiliary power available?

(6) Is auxiliary or replacement equipment available?

(7) Are open wires, terminal boxes, cross-connecting boxes, cables, and manholes frequently inspected for indications of sabotage and/or wire-tapping?

(8) Are maintenance crews alerted to search for tapping?

(9) Are civilian repairmen used? Are they checked and cleared?

(10) Have preparations been made to take care of sudden breaks in the system efficiently?

(11) Have personnel been cautioned about discussing classified or sensitive matters over unsecured telephone, teletype or radios?

b. Security weaknesses and recommendations.

(1) What specific weaknesses about the security of the communications system were noted during the survey?

(2) What specific and reasonable recommendations may be made to improve the security of the communications system?

17. Security of Information

a. Where on the installation are plans, blueprints, photos, classified material/ equipment, or other information of value to the enemy kept? The following list is not all-inclusive and is not a replacement for the checklist in OPNAVINST 5510.1.

(1) Is such material centralized in a single facility or scattered through various offices or buildings?

(2) In what sections is classified material processed stored and what level of classification is authorized in each area?

(3) Is all classified or valuable information kept in authorized/approved security containers or vaults?

(4) Are fight safes and cabinets affixed to floors or chained to immovable objects?

(5) Are container doors closed and locked when not in use?

(6) Is there any protection other than the container itself.

(7) What protection is given to a combination of containers?

(8) What security measures are enforced about keys to doors, gates, or file cabinets?

(9) Is access limited to combinations and keys?

(10) Who has access to combinations and keys? Do all authorized personnel have access? Have they been cautioned about passing keys and combinations to unauthorized personnel?

(11) Is a rigid chain of custody required for classified information (Secret and above)? Can custodians identify the location of classified at any time?

(12) Are only personnel with completed background checks and appropriate access assigned to positions requiring the handling of classified material?

(13) Are plans, blueprints, reports, or other classified material returned promptly and properly turned in?

(14) Who has access to classified material (with and without approved access)?

(15) Is dissemination of classified material strictly limited to those with a need to know?

(16) Is rank or position considered sufficient reason for access to classified information?

(17) Is classified material left unattended on desks where persons passing by can observe or steal without detection?

(18) Have civilian janitors been checked and placed under supervision?

(19) How is classified waste disposed of? Are destruction records kept?

(20) What policy has been established regarding releases and statements to local or national news media?

(21) Have all personnel been cautioned about unauthorized statements and releases?

b. Security of personnel.

(1) What specific weaknesses about the security of information were noted during the survey?

(2) What specific and reasonable recommendations may be made to improve the security of information?

18. <u>Security of Personnel</u>

a. OPNAVINST 5510.1 provides guidelines on personnel security.

b. Who is responsible for the security of the installation?

c. What is their attitude towards security?

d. Is the command aware of continuous evaluation of those who have access to classified or sensitive material or equipment?

e. Are personnel in positions of trust and confidence considered reliable?

f. What is their attitude towards security?

19. Identification System

a. What system is used to identify personnel authorized access within the confines of the installation/facility?

b. If badges are used—

(1) Are badges or identification cards of tamper-proof design and difficult to reproduce or counterfeit?

(2) Is the makeup and issue of badges and identification cards rigidly controlled to prevent:

(a) Reproduction?

(b) Theft?

(c) Unauthorized use or issue?

(d) Failure to return to issuing authority?

(3) Are photographs used on the face of the cards/badges?

(4) Is a detailed description used to positively identify holder?

(5) Are color or coded systems used to identify level of access for department personnel granted access?

(6) Are specific badges valid for specific areas?

(7) Is enforcement of such identification rigid?

(8) Do regulations prescribe that everyone wears the badge at all times and are regulations enforced?

(9) Is admittance to the installation/facility governed by the identification system?

(10) When badges are reported missing, lost, or forgotten what action is taken?

c. Is entrance permitted by wearing a military uniform?

(1) What other MEANS of identification are used?

(2) Are access rosters passed from one facility/command to another via secure means?

(3) Are passes or identification cards closely scrutinized?

d. What system is used to prevent persons working in one building, section, or unit from wandering about restricted areas without proper authorization?

20. Visitor Controls

a. What system is used to identify and admit authorized and legitimate visitors to the installation or facility?

(1) How and by whom is the legitimacy or necessity of a visitor's mission established?

(2) Are regulations lax in the control of visitors?

(3) On arrival at the gate, entrance of the facility or section, are visitors escorted to a reception area?

(4) Are regulations lax in the control of visitors?

(5) Is the identity of visitors verified?

(6) Is adequate information obtained from visitors?

(7) Is the purpose of the visit obtained?

(8) Are visitors required to register in a logbook with the following information?

(a) Full name.

(b) Social security number.

(c) Rank.

(d) Parent Organization.

(e) Date and time of entry.

(f) Time of departure.

(g) Number of security badge issued and level of access.

(h) Reason for visit.

(i) Name of official authorizing entry or providing escort.

(9) Are visitors required to provide identity on departure?

(10) Are visitors escorted or kept under surveillance during the time they are on the installation?

b. Is a vehicle register kept which includes:

(1) Date and time of entrance.

(2) Registration numbers.

(3) Name of owner(s).

(4) Signatures of driver(s) and passengers.

(5) Brief description of contents of vehicle.

(6) Inspections conducted on vehicle.

(7) Time of departure.

c. Are news media carefully checked and verified?

(1) Are credentials examined and verified?

(2) Has their visit been checked with higher commands to verify authority?

d. Are orders and credentials of allied military personnel examined by competent personnel (linguists, etc.).

(1) Are such visits verified by higher authority?

(2) Is security unduly sacrificed to courtesy?

e. Are spot checks of persons within the installation/facility made from time to time?

f. Security weaknesses and recommendations.

(1) What specific weaknesses about identification and visitor control were noted during the survey?

(2) What specific reasonable recommendations may be made to improve security by further identification and visitor control?

21. Description of Guard System

a. General description of the guard force.

(1) Strength.

(2) Shifts.

(3) Reserves.

(4) Weapons.

(5) Training.

(6) Number and type of posts.

(7) Communications.

b. Check the following points.

(1) What is the organization of the guard force?

(2) What is the numerical strength of each shift or relief?

(3) How many shifts or reliefs are there?

(4) How many supervisors does each shift have?

(5) Is supervision of the guard force adequate?

(6) How many fixed posts does the force cover?

(7) Where is each post located?

(8) How many patrols are covered by the guard force?

(9) What is the route of each patrol?

(10) Are routes of the patrols varied?

(11) What is the time of each patrol?

(12) Are doors and gates closely checked by the patrols?

(13) What functions are performed by the patrols?

(14) Does the supervisor make inspection tours of the routes?

(15) How frequently and thoroughly are such tours made?

(16) Are inspections varied as to route and time?

(17) Are guard force communications and alarm systems in use? Are they adequate?

(18) What type of communication and alarm system does the guard force have? Are there backup systems?

(19) Is a record kept of all guard force activity?

(20) Does the guard force have communications with the military police?

(21) What armament does the guard force have?

(22) Are the weapons in serviceable condition?

(23) Are the weapons suitable for the mission?

(24) Are arms and ammunition adequately safeguarded when not in use?

(25) Is there a record of custody when weapons are issued during each shift?

(26) Where are the weapons and ammunition stored? Does storage prevent rapid access to the guard force?

(27) How are guards recruited?

(28) What physical, mental, age, or other qualifications must protective guards have?

(29) How thoroughly are prospective guards investigated?

(30) Are guards uniformed, and do they have credentials or badges? What other system of identification is used?

(31) Is the guard force competent and respected by personnel of the installation?

(32) How thoroughly is the guard force trained?

(33) How much time is spent on training the guard force?

(34) How is the training of the guard force conducted?

(35) Does such training cover the following points?

(a) Care and use of weapons and ammunition.

(b) Common forms of espionage and sabotage activity.

(c) Common forms of bombs explosives.

(d) Familiarization with the installation/facility, with particular emphasis on restricted and vulnerable areas.

(e) Location and character of hazardous material and processes.

(f) Location and operation of important steam and gas valves and main electrical switches.

(g) Location and operation of fire protective equipment including use of sprinkler control valves.

(h) Conditions that may cause fires and explosions.

(i) Location and use of all first aid equipment.

(j) Duties in the event of fire, blackouts, or other emergencies that can be foreseen.

(k) Use of communication systems.

(l) Observation and description.

(m) Preservation of evidence.

(n) Patrol work.

(o) Searches of persons and places.

(p) Supervision of visitors.

(q) General and special guard orders.

(r) Location of all guard posts.

(36) Do guards have keys to gates, buildings, and offices?

(37) Do guards check the credentials of visitors and personnel working on the installation or facility?

(38) Is the strength of the guard force consistent with -

(a) Number of pedestrian, vehicle, and railroad gates and the hours they are open?

(b) Approximate number of daily visitors? Proper visitor reception?

(c) Number of loading platforms, storage facilities, working areas, etc.?

(d) Number of vehicles to cover the entire installation in a reasonable time?

(e) Number of restricted areas and vulnerable points?

(f) Number of plants or pumping stations?

(g) The number and extent of parking areas?

(h) Necessary supervision of the guard force?

(i) Sickness, leave, injury, etc., of guard personnel?

(39) What are the duties of the guard force if there are security violations? Does the guard force have security clearance and access?

c. Guard headquarters.

(1) Is the guard headquarters conveniently located?

(2) Is the guard headquarters properly secured at all times, and does it contain necessary equipment?

(3) Does the guard headquarters contain adequate facilities for members of the guard force?

d. Security weaknesses and recommendations.

(1) What specific security weaknesses were noted during the survey of the installation's guard force?

(2) What specific and reasonable recommendations regarding the guard force may be made to improve the security of the installation?

22. Description of Security Conditions and Security Measures of Adjacent Areas

a. What is the general nature of the population and the area surrounding the installation?

(1) Does the nationality or political nature of the surrounding populace offer a natural cover and aid to hostile agents and saboteurs?

(2) Is the installation within a commercial air zone of travel?

(3) If so, are minimum altitudes for planes published at all local airports?

(4) Is the installation isolated or screened from public view?

(5) Are restricted areas screened or isolated from public curiosity?

(6) Is the installation exposed to hazards brought onto the installation by natural conditions such as floods, extreme winds, forest fires, electrical storms, etc.?

(7) Is the installation or buildings within the installation well camouflaged against both air and ground observation?

(8) Have places of amusement near the installation and persons frequenting them been investigated, scrutinized, and checked?

(9) Have nightclubs, poolrooms, bowling alleys, houses of prostitution, barbershops, restaurants, taverns, stores, and other places frequented by personnel from the installation been included and thoroughly checked.

(10) Has the surrounding area been carefully scrutinized for any place likely to be used as bases for espionage or sabotage agents? Areas that could conceal antennas, audio and visual surveillance, etc.?

b. Security weaknesses and recommendations.

(1) What specific security weaknesses were noted during the survey of the area adjacent to the installation?

(2) What specific and reasonable recommendations may be made to improve security?

23. Security of Air Installations. The security of air installations does not differ from any other installation. Aircraft and maintenance facilities are high priority targets of saboteurs and espionage agents. In general, checking the following major areas will assist in establishing the security afforded to the installation.

a. Is the guard system adequate?

b. Are individual aircraft guarded sufficiently?

c. Are hangers and other vital buildings in a restricted area?

d. Have precautions been taken to see that there is no smoking in the area?

e. Are aircraft stored in hangers inspected periodically against sabotage?

f. Are special precautions taken to ensure visitor control in hangars?

g. Are vital repair parts in storage areas protected from unauthorized personnel, fire, and the elements?

h. Are there fire trucks, crash and rescue vehicles available?

i. Is emergency equipment parked in a convenient location readily available to any part of the installation?

24. Practical Use of Security Checklist

a. This checklist is not all encompassing and should be used as a guide to initiate a survey. Several methods of organizing a security check may be used. The following methods have been found to be practical and efficient.

(1) Itemize on index cards or automated data file requirements listed on the checklist and write the required information on each card/file as it is checked off the list.

(2) Itemize basic subdivisions of survey checklist requirements on separate pages with itemized requirements listed in required order. Write in the required information in the proper space as each item is checked off.

(3) Itemize all requirements of the survey checklist on separate pages, subdividing the pages according to main subdivision requirements. Make detailed notes about each item as it is checked off.

b. After completing notes on all requirements for each item, assemble in order and prepare report.

Section 14

Report of Investigative Activity

Purpose. Standard report used to report the results of a CI investigation.

CLASSIFICATION

Report of Investigative Activity

Record ID: ....

Point of Contact: ....

Classification: ....

Abstract: ....

Discretionary Access Control: ....

Caveats: ....

Release To: ....

Record Type: ....

Record Status: ....

Date Created (yyyymmdd): ....

Date Modified (yyyymmdd): ....

Community of Interest: ....

Source Record: ....

Case Number Reference: ....

Case: ....

Event: ....

Date of Record (yyyymmdd): ....

Title (Text): ....

Reason for Investigation (Text): ....

Individuals Involved (Text): ....

Role: ....

Status (Text): ....

Period of Report From (yyyymmdd): ....

Period of Report To (yyyymmdd): ....

Page number

CLASSIFICATION

CLASSIFICATION

Reporting Unit: ....

Agent Name: ....

Executive Summary Executive Summary (Text): ....

Page number

CLASSIFICATION

Section 15

Report of Investigative Activity Sworn Statement

Purpose. Standard report for preparing and documenting sworn statements.

CLASSIFICATION

Report of Investigative Activity Sworn Statement

Record ID: ....
Point of Contact:...
Classification: ....
Abstract: ....
Discretionary Access Control:....................................
Caveats: ....
Release To: ....
Record Type: ....
Record Status:..
Date Created (yyyymmdd): ....
Date Modified (yyyymmdd):..
Community of Interest:..
Source Record: ....
Report Detail Date (yyyymmdd): ....
Case Number: ....
Lead Number:..
ROIA Number: ....
Name: ....
Title..
Sub Title ....
Agency:..
Number of Investigative Materials:.............................
Text..
Agent Name:..
Organization: ....
Individual Information: ....
Name: ....

Page number

CLASSIFICATION

CLASSIFICATION

Employer:..

OR Unit:...

Sworn Statement ....

Number of Witnesses:..

Page number

CLASSIFICATION

Section 16

Personal Data Form—POW/MIA/MISSING (Non-Hostile)

Purpose. Standard report used to record and document POW/MIA/missing personnel investigations.

CLASSIFICATION

Personal Data Form—POW/MIA/MISSING (Non-Hostile)

1. Personal Data

   a. Name:

   b. Rank:

   c. SSN/MOS:

   d. Former Service Number:

   e. Organization:

   f. Date of Birth:

   g. Place of Birth:

   h. Home of Record:

   i. Residence (if other than home of record):

   j. Marital Status (Include number, sex, citizen status, and age of children):

   k. PEBD:

   l. EAS/EOS:

   m. Date arrived in country:

   n. Duty assignment:

2. Physical Characteristics

   a. Height (Metric as well as U.S. equivalent):

   b. Weight (Metric as well as U.S. equivalent):

Page number

CLASSIFICATION

c. Build:

d. Hair:

e. Eyes:

f. Complexion:

g. Race:

h. Right/left handed:

3. Distinguishing Characteristics

a. Speech (Include accent and speech patterns used):

b. Mannerisms:

c. Scars/identifying marks (Include type, location, size, color, and detailed description):

d. Others:

4. Circumstances of Incident

a. Date:

b. Location (Coordinates and geographic name):

c. Circumstances:

d. Reported wounds:

e. Last known location:

f. Last known direction of travel:

g. Last known place of detention:

h. Status (prisoner of war/missing [non-hostile]/missing in action as reported by unit):

5. Other Pertinent Data

a. General physical condition:

b. Linguistic capabilities and fluency:

Page number

CLASSIFICATION

c. Religion:

d. Civilian education:

e. Military schools:

f. Clothing and equipment when last seen:

g. Jewelry when last seen (Include description of glasses, rings, watches, religious medallions, etc.):

h. Other personnel listed POW/MIA during same incident:

6. Photograph

7. Handwriting Samples (Attach sample of correspondence, notes, etc. If no other sample is available, include reproduction of signature from Service Record Book/ Officer's Qualification Record)

Enclosures: (May not be given wide dissemination based on classification or content.)

a. Clearances/Access Information. (Include information concerning security clearance, access, knowledge of recurring tactical operations, knowledge of projected or proposed operations, or any other special knowledge possessed.)

b. Medical Profile. (Include pertinent information extracted from medical records and summarized information gained concerning ability to survive in captivity, known personal problems, relationship with seniors/contemporaries or other personal, medical, or personality information which would indicate his ability to cope in a prisoner-of-war situation.)

c. References. (List any messages, letters, or other correspondence pertaining to the individual. If circumstances under which the individual is listed as captured or missing predicated a command investigation, a copy of that investigation is included as an enclosure.)

d. Unresolved Leads/Investigators Comments. (Include unresolved leads or names of personnel who were unavailable for interview because of transfer, evacuation, etc. Use investigator's comments as necessary but do not recommend a casualty determination.)

This page intentionally left blank.

Section 17

Counterintelligence Measures Worksheet

CLASSIFICATION

Counterintelligence Measures Worksheet

Purpose. The CI measures worksheet (see page D-64) is prepared or revised based on the conclusions reached in the intelligence estimate of the enemy capabilities for intelligence, subversion, terrorist activities, and sabotage. This worksheet is an essential aid in CI planning. It is also the basic for preparing CI orders and requests. The following is a partially completed sample of a CI measures worksheet.

Page number

CLASSIFICATION

(1) Phases or Periods of the Operation	(2) Categories of Counter-intelligence Activities Involved	(3) Counterintelligence Measures to be Adopted	(4) Units/Personnel Responsible for Execution of CI Measures						(5) Instructions Regarding Entries in Columns (3) and (4). Notes for Future Action, and Staff Coordination Measures
			Civil Affairs	CIS Officer	Provost Marshal	Comm Intel Units	All Units	CI Units	
Assault Phase	1. Military Security a. Security discipline	(1) Cover or paint all vehicle and aircraft markings.					X		Coordinate with G-4 Provost marshal report violations Coordinate with G-1 Provost marshal report violation Coordinate with G-3 CI elements assist with instruction and check SOP Coordinate with G-1 SOP CI elements check SOP CI elements check SOP Check with CIS officer for compliance Coordinate with other military forces Coordinate with G-3 and G-4 Provost marshal report violation
		(2) Remove identification from uniform.					X		
		(3) Restrict personnel to area except when on official business.					X		
		(4) Emphasize security discipline in command posts/echelons, and elsewhere, with particular reference to handling of communications and information systems, documents and maps, phone conversations, loose talk, and speculation which might convey information to the enemy. All personnel will be instructed regarding same.					X		
		(5) Report all known or suspected security compromises to unit security manager and intelligence officer.		X	X	X	X	X	
	b. Safe-guarding of classified information and equipment	(6) Collect and place under guard or evacuate as determined appropriate by unit commander civilians in position to observe critical unit C2, fires, and combat service support sites.	X					X	
	c. Communication and information security	(7) Check SOP plans for security of cryptographic devices for destruction and for report of loss or compromise.		X	X	X	X	X	
		(8) Check plans and equipment for destruction of documents in event of imminent capture.		X			X	X	
	d. Security of unit movements	(9) Use only authorized call signs, authenticators, and crytographic codes.		X			X		
		(10) Check that unauthorized personnel are prohibited from entering CPs, message centers, and other sensitive areas.		X		X	X	X	
		(11) Patrol all wire lines used by units.		X			X		
		(12) Cut all wire lines leading into enemy-occupied territory.		X			X		
		(13) Control the movement of all vehicles and aircraft to the extent that a change in normal operations is not indicated.					X		
		(Others as required)							

Page number

CLASSIFICATION

Appendix E. Counterintelligence Training Courses

MAGTF CI BASIC COURSE (U)

What This Course Offers

Provides instruction in theater, national, DOD, and organic Marine Corps intelligence assets; CI application of the combat intelligence cycle; CI hostile threat; terrorism; CI/tactical HUMINT operations; photography; interrogations; espionage, sabotage, subversion, and terrorism investigations; interview skills; intelligence report writing; and surveillance techniques.

Description

Trains USMC enlisted and officers serving as members of a CI team, or subteam, in support of a MAGTF .

Emphasis is placed on requirements in amphibious and subsequent operations.

Trains USMC enlisted and officers in CI/HUMINT related tasks when serving as a member of a CI team, CI/HUMINT branch, or in support of MAGTF/JTF command.

Emphasizes theatre and national-level CI support provided to the commander.

Who Should Attend

Marine Corps corporal through lieutenant screened by CI assets and approved for lateral move to MOS 0211/0210/0204 by HQMC in accordance with MCO 3850.1H. Other services (for example, U.S. Army enlisted) personnel have attended. Projected attendance for additional U.S. Army and possible U.S. Air Force personnel is anticipated.

Course Activities

Lectures, videos, discussions, and practical exercises

Faculty

Navy and Marine Corps Intelligence Training Center, Dam Neck, VA

How Long? How Often?

Seventeen and a half weeks, 88 training days/annually

Security Clearance Needed

Top Secret based on completed SSBI and eligible for SCI access

Further Information

HQ, US Marine Corps, code IOC at (703) 614-2219/2058, DSN 224-XXXX

MAGTF ADVANCED CI COURSE (U)

What This Course Offers

Provides instruction on CI/Tactical HUMINT collection; intelligence architecture; systems and communications; MAGTF, theater, and national-level staff planning; MAGTF/JTF/CI/ITT employment and deployment;

case method leadership practicums; CI espionage, sabotage, subversion, and terrorism theory; and legalities of investigations.

Who Should Attend

Marine Corps Gunnery Sergeant through Lieutenant Colonel (MOS 0211/0210/0204) with at least one successful tour. Other service quotas are available.

Course Activities

Lectures, videos, discussions, and practical exercises

Faculty

Navy and Marine Corps Intelligence Training Center, Dam Neck, Virginia

How Long? How Often?

Twenty-six days, 20 training days/annually

Security Clearance Needed

Top Secret/SCI

Further Information

HQMC, code IOC (703) 614-2219/2058, DSN 224-XXXX

ADVANCED FOREIGN CI TRAINING COURSE (U)

Description

Provides tough, demanding, realistic, mission focused counterespionage training to selected foreign CI special agents who are programmed for assignment within the national foreign CI community in support of the warfighters and the DOD foreign CI strategy. Provides students with a synergistic perspective for applying advanced foreign CI skills and methodologies focusing on the foreign CI Triad of investigations, operations, and surveillance.

What This Course Offers

Advanced counterespionage concepts, principles, and techniques for foreign CI special agents

Who Should Attend

DOD CI special agents with three years of strategic CI experience; 12 seats per class for Army and two seats designated joint, which rotate among the military services.

Course Activities

Lectures, presentations, discussions, case studies, practical exercises, and videos

Faculty

Eight full-time instructors with intensive counterespionage background supplemented with guest speakers/subject-matter experts from DOD and other Intelligence Community components.

How Long? How Often?

Fifteen weeks (approximately 750 hours)/offered twice a year

Security Clearance Needed

Top Secret

Further Information

Contact the course director or senior instructor at (301) 677-5778/5779, FAX (301) 677-6362. Mailing address is Commander, U.S. Army Foreign CI Activity (USAFCA), USAINSCOM, Attn. IAFC-TC, Fort Meade, MD 20755

Registration Data

Limited to those working in or en route to a foreign CI assignment; graduate of a basic CI course; three years of strategic CI experience; effective communicator; supervisory and command recommendations; favorable SSBI; CI-scope polygraph; and a valid civilian drivers license. Army registration procedures, a special nomination packet must be submitted to Commander, U.S. Army Intelligence and Security Command; Attn. IAOPS-HUCI, Fort Belvoir, VA 22060-5246. An INSCOM selection board chooses students on a best qualified basis. Other military services per service directives and guidance.

CI ANALYTIC METHODS COURSE (U)

Description

Introduction to multi-discipline CI analytical methods; tools; matrix, link, and pattern analysis; collection threats; deception analysis; and intelligence integration.

Who Should Attend

Entry-level CI analysts

Course Activities

Lectures, discussions, videos, and case studies

Faculty

Instructors from the JMTC

How Long? How Often?

One week/two times a year

Security Clearance Needed

Top Secret/SCI

Further Information

JMTC, DIAC, Bolling AFB, (202) 373-3312

JOINT CI STAFF OFFICERS COURSE (U)

Description

Know how CI activities are integrated into joint-military organizations at varioius levels of command and into the formulation of contigency plans.

Know how to develop and execute a CI appendix to the intelligence annex to a joint operational/exercise plan or OP order.

Know the Joint Planning System and how it supports both deliberate and time-sensitive plans.

Know the roles/responsibilities of DOD, NSA, FBI, CIA, JCS, combatant commands, and the Services CI agencies in providing CI support to contigency planning and execution.

Know how CI HUMINT and Special Operations Forces will coordinate and deconflict activities in a contingency plan.

What This Course Offers

This course introduces the student to CI support to joint operations.

Who Should Attend

Personnel who will be working in a joint CI support role during contingencies.

Course Activities

Lectures, videos, discussions, and practical exercises

Faculty

JCISB and civilian and military members of the intelligence and CI communities

How Long? How Often?

Five days/several times a year

Security Clearance Needed

Top Secret/SCI

Further Information

DIA (DAC-1B), Joint CI Support Branch (JCISB), Pentagon, Room 1E821, (703) 614-9155.

MULTI-DISCIPLINE CI COURSE (U)

What This Course Offers

Improves professional CI officers' understanding of the multi-discipline approach to CI. This is not a course in analytic methods or methodology.

Who Should Attend

CI professionals from throughout DOD and the Intelligence Community (IC).

Course Activities

Lectures, discussions, videos, and case studies

Faculty

Instructors from the JMTC and subject-matter experts from the intelligence community

How Long? How Often?

Two weeks/three times a year/also available as a two-three day mobile course.

Security Clearance Needed

Top Secret/SI/TK/G

Further Information

JMTC, DIAC, Bolling AFB, Washington, D.C. (202) 373-3897

Description

Learn the all-source CI national and DOD environments.

Learn how to access HUMINT, SIGINT, IMINT, MASINT, and other information and resources in the IC.

Learn how organizationally perceived roles affect CI policy and analysis.

Learn the threats to U.S. national interests from foreign intelligence and security services (FISS) and about the U.S. resources that drive responses to FISS threats.

Understand the complex interdependent relationships of CI organizations from operations to finished analytical production.

EVOLUTION OF AMERICAN CI (U)

What This Course Offers

Provides the students with a broad historical perspective of the growth and development of U.S. CI from the historical legacy of the American Revolution to the current day.

Who Should Attend

New or mid-level intelligence officers or special agents whose present or future assignments may involve CI responsibilities.

Course Activities

Lectures, videos, case studies, and class participation

Faculty

NACIC and guest speakers

How Long? How Often?

One week/twice a year

Description

Evaluates the historical significance of key events in the development of the CI discipline in the United States.

Applies the lessons learned to future public and legislative scrutiny and helps students make decisions based on a historical perspective.

Projects the need for a strong CI program.

Lists and explains the five core issues affecting CI policies and strategies in the post-Cold War era.

Advances interagency cooperation and creates a learning environment.

Identifies various agencies' perspectives and stimulates creative thinking.

Security Clearance Needed

Secret

Further Information

Community Training Branch, NACIC, (703) 874-4122

Registration Data

Three weeks before course

STRATEGIC APPROACHES TO CI (SACI) (U)

Description

Analyzes foreign intelligence services CI threat data.

Determines risks and vulnerabilities to information.

What This Course Offers

Designed to illuminate the "big picture," emphasizes U.S. national CI strategies and the approaches taken throughout the CI community to implement these strategies. Focuses on the five core issues that will be enduring challenges for the CI community for the next decade.

Who Should Attend

New or mid-level managers in operational CI agencies or other CI community elements who demonstrate a potential for advancement to more senior CI positions, and whose present or future assignments may involve CI policy formation or interagency responsibilities.

Course Activities

Lectures, videos, case studies, group discussions (to develop a strategy—with examples of successful strategies used by U.S. companies to compete in today's world), student presentations.

Faculty

Instructors, program managers, and subject-matter experts from throughout the intelligence and CI communities, sponsored by the NACIC.

How Long? How Often? Class Size

Fifty hours/April and October/30

Security Clearance Needed

Top Secret/SCI

Further Information

Community Training Branch, NACIC, (703) 874-4122. Usually at an off-site location.

Registration Data

Nomination by parent agency. Thirty days before course date.

THE THREAT TO INFORMATION SYSTEMS (U)

What This Course Offers

This course reviews an updated summary of various threats to information systems.

Who Should Attend

Students should have familiarity with the concept of threats from foreign intelligence services

Course Activities

Lectures, discussions, demonstrations, and practical problems

Faculty

Instructors from the National Cryptologic School (NCS)

How Long?

Twenty-four hours/three days, full-time

Security Clearance Needed

Top Secret and indoctrinated for special intelligence. Pass to NSA Office of Security electronically.

Further Information

NCS (410) 859-6336 or secure 968-8054

Registration Data

Open to CI community personnel. Registration request should be made through parent agency training coordinator to the NSA/NCS registrar.

INFOSEC FAMILIARIZATION COURSE (U)

What This Course Offers

This course is a survey of communications security (COMSEC) principles and techniques with an emphasis on electronic COMSEC systems and cryptographic equipment.

Description

Topics include—

- A history of COMSEC and cryptology.
- The national information security (INFOSEC) structure, mission, and relationships.
- The vulnerability of threats to U.S. military and civil communications systems.
- Physical, cryptographic, transmissions, and emission (TEMPEST) security.
- Off-line cryptosystems.
- Emergency destruction.
- COMSEC material production.
- Computer security digital encryption theory.
- Key management.
- INFOSEC system and cryptographic equipment applications.
- Systems and equipment under development.
- INFOSEC trends.

Who Should Attend

Students requiring fundamental knowledge of COMSEC

Course Activities

Lectures and discussions

Faculty

Instructors from the National Cryptologic School (NCS)

How Long?

Forty hours/one week, full-time

Security Clearance Needed

Top Secret. Pass to NSA Office of Security electronically.

Further Information

NCS (410) 859-6336 or secure 968-8054

Registration Data

Open to CI community personnel. Registration request should be made through parent agency training coordinator to the NSA/NCS registrar.

Appendix F. MAGTF Counterintelligence Planning Checklist

This appendix identifies typical MAGTF CI/HUMINT planning tasks and activities during each phase of the MCPP. Most planning tasks and activities require the coordinated action of various MAGTF G-2/S-2 section and intel bn personnel.

MCPP	Actions	Counterintelligence Planning Actions
Mission Analysis	Identify the higher headquarter's (HHQ) supported headquarters intent. Identify tasks. Determine the AO and area of interest (AOI). Review available assets and identify personnel and equipment resource shortfalls. Determine constraints and restraints. Determine recommended commander's critical information requirements (priority intelligence requirements, friendly force information requirements, EEFI). Identify requests for information. Determine assumptions. Draft mission statement. Present mission analysis brief. Draft the warning order. Convene/alert red cell (if appropriate). Begin staff estimates. Refine commander's intent Develop the commander's planning guidance.	Review HHQ and MAGTF standing intelligence plans (e.g., Annex B to an OPLAN), CI plans (Appendix 3 to Annex B), HUMINT plan (Appendix 5 to Annex B), etc. Assist with determination of the MAGTF AO and AOI. Assess DIA's, CIA's, combatant commands, and other external organizations ongoing CI operations and plans within the AO and AOI (e.g., availability and currency of CI contingency materials). Provide initial CI estimates and other CI products to support initial planning (ensure needs of subordinate units are identified and met). Determine specified, implied, and essential CI tasks. Develop proposed CI mission statement; coordinate with G-2/S-2 plans officer, the ISC, the CI/HUMINT Co commander, and the G-3/S-3 force protection officer; obtain G-2/S-2 officer's approval. Assist security manager with development of security classification guidance to support planning and subsequent operations Identify organic/supporting CI elements & subordinate units' CI points of contact; acquire an immediate operational status report from each; determine personnel and equipment deficiencies. Review/prepare new CI survey/vulnerability assessment; determine and prioritize significant security vulnerabilities; provide G-3/S-3 recommendations (e.g., CI active and passive measures); identify requirements for technical surveillance countermeasures support. Identify JTF/multinational CI interoperability issues; provide recommendations. Establish/review/update the MAGTF CI data bases; special attention to current threat estimates, current CI estimates, and CI targets (personalities, organizations, and installations). Ensure subordinate units CI POCs kept advised of pertinent actions and developments. Identify external organizations CI collection, production, and dissemination plans, and assess against MAGTF's initial operations requirements and plans. Determine CI personnel & equipment deficiencies; initiate augmentation requests (coordinate with intel bn commander and the intelligence operations officer). Assign/task-organize organic CI elements (e.g., CI/HUMINT company detachments or HUMINT support teams to major subordinate elements; CI element to MAGTF future operations/plans sections); ensure that detailed C2 relationships, authorities, and restrictions are prepared and disseminated to all concerned. Validate/update JTF CI tactics, techniques, and procedures and MAGTF SOP (coordinate with HHQ and subordinate units). Validate and prioritize CI requirements; special attention to those needed for COA development. Begin development of CI operations plan; issue orders to CI collection, production, and dissemination elements (coordinate with ISC, G-2 plans and operations officers, CMDO, P&A cell OIC, and SARC OIC). Determine initial CI CIS requirements and dissemination plans; identify deficiencies (coordinate with ISC, G-2 plans and operations officers, CMDO, and the G-6/S-6). Validate CI data base management procedures (coordinate with P&A cell OIC, CI/HUMINT Co commander, JTF and subordinate units). Keep subordinate units' CI POCs advised of pertinent actions and developments. Determine/begin development of CI criminal investigation authorities and relationships with NCIS and PMO.

MCPP	Actions	Counterintelligence Planning Actions
COA Development	Continue intelligence preparation of the battlespace (throughout all steps of the planning process). Array friendly forces. Assess relative combat power. Conduct centers of gravity and critical vulnerabilities analysis. Brainstorm possibilities. Develop rough cut COA. Commander's input. Refine COA(s). Validate COA(s). Develop COA(s) graphic and narrative. Prepare and present COA(s) briefing. Commander selects/ modifies COA(s).	Assist with development and continued updating of the intelligence and CI estimates, with emphasis on the following: • CI target reduction plans development. • Periodic CI summaries and threat estimate update. Recommendations and implementation of current/future CI countermeasures. Assist the intelligence, operations, and other staff sections with COA development. Develop the CI concept of operations for each COA; begin preparation of— • Appendix 3 (CI operations) to Annex B • Assistance to Appendix 5 (HUMINT operations) to Annex B • Assist G-3/S-3 section with force protection plans to Annex C Determine CI capabilities required for each COA. Identify and coordinate CI-related collection, production, and dissemination requirements for each COA. Continue development of CI estimate of supportability for each COA. Ensure subordinate units' CI POCs kept advised of pertinent actions and developments.
COA Analysis	Conduct COA analysis wargaming. Refine staff estimates and estimates of supportability. Develop concepts based upon warfighting functions (as required). Prepare COA analysis brief.	Complete CI estimate and threat assessments. Complete CI estimates of supportability. Assist G-2/S-2 section with completion of the intelligence estimate and the friendly intelligence estimate of supportability. Assist G-6/S-6 section with completion of the force protection estimate. Continue to monitor and update CI collection, production, and dissemination activities. Ensure subordinate units receive necessary CI products; verify understanding; and identify/update subordinates' current IR and force protection EEFIs. Validate and update CI IRs. Ensure subordinate units CI POCs kept advised of pertinent actions and developments.

MCPP	Actions	Counterintelligence Planning Actions
COA Comparison and Decision	Evaluate each COA. Compare COAs. Commander's decision. Issue warning order.	Assist G-2/S-2 and G-3/S-3 sections with evaluation and comparison of each COA. Continue development of appendix 3 to Annex B consistent with the selected COA. Update, validate and prioritize CI IRs and supporting CI collection/production requirements for the selected COA; issue orders as appropriate to CI elements. Coordinate CI element task-organization needs associated with the selected COA, with special attention to necessary support to the main effort. Update/develop in detail supporting C2 relationships, authorities, and restrictions. Continue coordination with the G-6/S-6 regarding CI CIS requirements, to include standard and unique CIS for internal CI operations and with other joint/multinational organizations. Continue coordination with G-1/S-1 as necessary for physical couriering of CI products to subordinate units; and with the G-1/S-1 and PMO for EPW handling/compound related plans development. Review actions associated with satisfying CI personnel and equipment deficiencies associated with the selected COA. Ensure subordinate units receive pertinent CI products (e.g., current CI threat assessment); verify understanding; identify/update subordinates current CI-related intelligence requirements and EEFIs. Validate MAGTF CI-related intelligence requirements and tasks to support force protection EEFIs. Ensure subordinate units CI POCs kept advised of pertinent actions and developments.
Orders Development	Refine commander's intent. Turn concept of operations into an operations order or a fragmentary order. Update and convert staff estimates and other planning documents into OPORD annexes and appendices. Commander approves OPORD.	Complete appendix 3 to Annex B; ensure copies provided to subordinate units and they understand it. Assist with completion of Appendix 16 (Intelligence Operations Plan) and other appendices to Annex B. Update, validate and prioritize CI IRs and associated collection, production, and dissemination operations. Monitor ongoing CI production operations, update and issue orders as appropriate to CI elements. Ensure pertinent CI products are disseminated to subordinate units. Update/finalize CI criminal investigation plans with NCIS and PMO. Complete CI related CIS actions. Maintain coordination with external CI elements.
Transition	Transition brief. Drills. Plan refinements (as required).	Assist intelligence section with transition brief. Modify CI plans as necessary. Monitor ongoing CI collection and production operations; update and issue orders as needed to CI elements. Ensure subordinate units' CI POCs and CI officers in JTF and other components fully understand plans and standing requirements; and ensure they have received necessary CI products. Identify, validate, and prioritize remaining CI IRs and force protection EEFIs. Participate in drills. Remain engaged in MAGTF future plans activities.

Appendix G. Glossary

Section I. Acronyms

ACE . . . . . . aviation combat element
AFP . . . . . . all-source fusion platoon
AOR . . . . . . area of responsibility
ARG . . . . . . amphibious ready group
ASAS . . . . . . all source analysis system (Army)
ATFIC . . . . . . amphibious task force intelligence center

BDA . . . . . . battle damage assessment
bn. . . . . . . battalion

C2 . . . . . . command and control
C2W . . . . . . command and control warfare
CA. . . . . . civil affairs
CAP. . . . . . crisis action planning
CE . . . . . . command element
C-E . . . . . . communications-electronic
CFSO. . . . . counterintelligence force protection source operations
C-HUMINT. . . . . . counter-human intelligence
CHAT . . . . . . CI/HUMINT automated tool set
CI. . . . . . counterintelligence
CICM . . . . . . counterintelligence contingency material
CIC . . . . . . combat intelligence center
CID . . . . . . criminal investigation division
CIHEP. . . . . . CI/HUMINT equipment program
CIHO. . . . . . CI/HUMINT officer
C-IMINT. . . . . . counter imagery intelligence
CINC. . . . . . commander in chief
CIS . . . . . . communications and information systems
CISUM . . . . . . counterintelligence summary
CITEX. . . . . . counterintelligence training exercise
CLF. . . . . . commander, landing force
CMD . . . . . . collection management and dissemination
CMDO. . . . . . collection management and dissemination officer
co. . . . . . company
COA . . . . . . course of action
COC . . . . . . current operations center
CP . . . . . . command post
CPX. . . . . . command post exercise
C-SIGINT . . . . . . counter signals intelligence
CSS . . . . . . combat service support
CSSE. . . . . . combat service support element

DCID. . . . . . Director of Central Intelligence Directive
DCISS . . . . . . Defense Counterintelligence Information System
DES . . . . . . digital encryption loader
det . . . . . . detachment
DHS. . . . . . Defense HUMINT Service
DIA . . . . . . Defense Intelligence Agency
DIAM. . . . . . Defense Intelligence Agency manual
DITDS . . . . . Defense Intelligence Threat Data Systems
DOD. . . . . . Department of Defense
DON. . . . . . Department of the Navy
DS . . . . . . direct support
DST . . . . . . direct support team

EA . . . . . . electronic attack
EEFI. . . . . . essential elements of friendly information
EPW. . . . . . enemy prisoner of war
ES. . . . . . electronic support
EW. . . . . . electronic warfare

FCIP. . . . . . foreign counterintelligence program
FFCC . . . . . . force fires coordination center
FIS . . . . . . foreign intelligence services
FISS . . . . . . foreign intelligence and security service
FOC . . . . . . future operations center
FORMICA . . . . foreign military intelligence collection activity
FSCC . . . . . . fire support coordination center

GCE . . . . . . ground combat element
GS . . . . . . general support
GSP . . . . . . ground sensor platoon

HLZ . . . . . . helicopter landing zone
HMMWV. . . . . . high mobility multipurpose wheeled vehicle
HOC. . . . . . HUMINT operations cell
HOCNET . . . . HUMINT Operational Communications Network
HST . . . . . . HUMINT support team
HUMINT . . . . . . human intelligence

I&W. . . . . . indications and warning
IAS. . . . . . intelligence analysis system
ICR. . . . . . intelligence collection requirement
IDR. . . . . . intelligence dissemination requirement
IHR. . . . . . in extremis hostage rescue
IIP. . . . . . imagery intelligence platoon
IMINT . . . . . . imagery intelligence
INFOSEC. . . . . . information security
INTEL . . . . . . intelligence
IO . . . . . . information operations
IOC. . . . . . intelligence operations center
IPB . . . . . . intelligence preparation of the battlespace
IPR . . . . . . intelligence production requirement
IR . . . . . . intelligence requirement

ISC . . . . intelligence support coordinator
ISOPREP . . . . isolated personnel report
IT . . . . interrogator-translator
J-2. . . . . joint staff/force intelligence officer
J-2X . . . . joint force J-2 CI/HUMINT staff element
JCISB . . . . joint CI support branch
JDISS . . . Joint Deployable Intelligence Support System
JFC . . . . joint force commander
JIC . . . . joint intelligence center
JIDC . . . . joint interrogation and debriefing center
JISE . . . . joint intelligence support element
JOPES . . . . Joint Operation Planning and Execution System
JTF . . . . joint task force
JWICS . . . . Joint Worldwide Intelligence Communications System

LAN . . . . local area network

MAG . . . . Marine aircraft group
MAGTF . . . . Marine air-ground task force
MARFOR . . . . Marine Corps Forces
MARFORLANT . . . . . . Marine Corps Forces, Atlantic
MARFORPAC . . . . Marine Corps Forces, Pacific
MARFORRES . . . . Marine Corps Forces Reserve
MASINT . . . . . measurement and signature intelligence
MCIA . . . . Marine Corps Intelligence Activity
MCISU . . . . Marine Corps Imagery Support Unit
MCPP . . . . Marine Corps Planning Process
MDITDS . . . . Migration Defense Intelligence Threat Data System
MEF . . . . Marine expeditionary force
METT-T . . . . mission, enemy, terrain and weather, troops and support available-time available
MEU(SOC) . . . . Marine expeditionary unit (special operations capable)
MHG . . . . MEF headquarters group
MO . . . . modus operandi
MOOTW . . . . military operations other than war
MOS . . . . military occupational specialty
MOTC . . . . Military Operations Training Course
MP . . . . military police
MSC . . . . major subordinate command

NCIS . . . . Naval Criminal Investigative Service
NGO . . . . nongovernmental organization
NIPRNET . . nonsecure internet protocol router network
NMITC . . . . Navy Marine Corps Intelligence Training Center
NSA . . . . National Security Agency

obj . . . . objective
OCAC . . . . operations control and analysis center
OFCO . . . . offensive counterintelligence operations
OIC . . . . officer in charge
OMFTS . . . . operational maneuver from the sea
OOB . . . . order of battle
OPCON . . . . operational control
OPLAN . . . . operation plan
OPORD . . . . operation order
OPSEC . . . . operations security
OSI . . . . . . Office of Special Investigations (Air Force)
OSINT . . . . open-source intelligence

P&A . . . . production and analysis
PIR . . . . priority intelligence requirement
plt . . . . platoon
PMO . . . . Provost Marshal office
POI . . . . personalities, organizations, and installations
POI&I . . . . personalities, organizations, installations, and incidents
PSYOP . . . . psychological operations

QSTAG . . . . . . quadripartite standardization agreement

RFC . . . . raid force commander
ROE . . . . rules of engagement

SARC . . . . surveillance and reconnaissance cell
SASO . . . . stabilization and security operations
SCI . . . . sensitive compartmented information
SERE . . . . survival, evasion, resistance, escape
SIGINT . . . . signals intelligence
SIPRNET . . SECRET internet protocol router network
SLAR . . . . side looking airborne radar
SOA . . . . sustained operations ashore
SODARS . . . . special operations debrief and retrieval system
SOFA . . . . status-of-forces agreement
SOP . . . . standing operating procedures
SPMAGTF . . . . special purpose Marine air-ground task force

TACON . . . . tactical control
TDN . . . . tactical data network
T/E . . . . table of equipment
TFCICA . . . . task force CI coordinating authority
TIARA . . . . . tactical intelligence and related activities
TM . . . . technical manual
topo . . . . topographic
TPFDD . . . . . . time-phased force and deployment data
TRAP . . . . . . tactical recovery of aircraft and personnel
TSCM . . . . technical surveillance countermeasures
TTP . . . . tactics, techniques, and procedures

UAV . . . . unmanned aerial vehicle

WAN . . . . wide area network

Section II—Definitions

accountability—The obligation imposed by law or lawful order or regulation on an officer or other person for keeping accurate record of property, documents, or funds. The person having this obligation may or may not have actual possession of the property, documents, or funds. Accountability is concerned primarily with records, while responsibility is concerned primarily with custody, care, and safekeeping. (JP 1-02)

administrative control—Direction or exercise of authority over subordinate or other organizations in respect to administration and support, including organization of Service forces, control of resources and equipment, personnel management, unit logistics, individual and unit training, readiness, mobilization, demobilization, discipline, and other matters not included in the operational missions of the subordinate or other organizations. Also called ADCON. (JP 1-02)

agent—In intelligence usage, one who is authorized or instructed to obtain or to assist in obtaining information for intelligence or CI purposes. (JP 1-02)

agent authentication—The technical support task of providing an agent with personal documents, accoutrements, and equipment which have the appearance of authenticity as to claimed origin and which support and are consistent with the agent's cover story. (JP 1-02)

agent net—An organization for clandestine purposes which operates under the direction of a principal agent. (JP 1-02)

all-source intelligence—Intelligence products and/or organizations and activities that incorporate all sources of information, including, most frequently, human resources intelligence, imagery intelligence, measurement and signature intelligence, signals intelligence, and open source data, in the production of finished intelligence. (JP 1-02)

antiterrorism—Defensive measures used to reduce the vulnerability of individuals and property to terrorist acts, to include limited response and containment by local military forces. (JP 1-02)

area of interest—That area of concern to the commander, including the area of influence, areas adjacent thereto, and extending into enemy territory to the objectives of current or planned operations. This area also includes areas occupied by enemy forces who could jeopardize the accomplishment of the mission. Also called AOI. (JP 1-02)

area of operation—An operational area defined by the joint force commander for land and naval forces. Areas of operation do not typically encompass the entire operational area of the joint force commander, but should be large enough for component commanders to accomplish their missions and protect their forces. Also called AO. (JP 1-02)

assessment—**1.** Analysis of the security, effectiveness, and potential of an existing or planned intelligence activity. **2.** Judgment of the motives, qualifications, and characteristics of present or prospective employees or "agents." (JP 1-02)

asset (intelligence)—Any resource -person, group, relationship, instrument, installation, or supply -at the disposition of an intelligence organization for use in an operational or support role. Often used with a qualifying term such as agent asset or propaganda asset. (JP 1-02)

assign—**1.** To place units or personnel in an organization where such placement is relatively permanent, and/or where such organization controls and administers the units or personnel for the primary function, or greater portion of the functions, of the unit or personnel. **2.** To detail individuals to specific duties or functions where such duties or functions are primary and/or relatively permanent. (JP 1-02)

attach—**1.** The placement of units or personnel in an organization where such placement is relatively temporary. **2.** The detailing of individuals to specific functions where such functions are secondary or relatively temporary, e.g., attached for quarters and rations; attached for flying duty. (JP 1-02)

aviation combat element—The core element of a Marine air-ground task force that is task-organized to conduct aviation operations. The aviation combat element provides all or a portion of the six functions of Marine aviation necessary to accomplish the Marine air-ground task force's mission. These functions are antiair warfare, offensive air support, assault support, electronic warfare, air reconnaissance, and control of aircraft and missiles. The aviation combat element is usually composed of an aviation unit headquarters and various other aviation units or their detachments. It can vary in size from a small aviation detachment of specifically required aircraft to one or more Marine aircraft wings. The aviation combat element may contain other Service or foreign military forces assigned or attached to the Marine air-ground task force. The aviation combat element itself is not a formal command. Also called ACE. (Approved for inclusion in next version of MCRP 5-12C)

basic intelligence—Fundamental intelligence concerning the general situation, resources, capabilities, and vulnerabilities of foreign countries or areas which may be used as reference material in the planning of operations at any level and in evaluating subsequent information relating to the same subject. (JP 1-02)

battle damage assessment—The timely and accurate estimate of damage resulting from the application of military force, either lethal or non-lethal, against a predetermined objective. Battle damage assessment can be applied to the employment of all types of weapon systems (air, ground, naval, and special forces weapon systems) throughout the range of military operations. Battle damage assessment is primarily an intelligence responsibility with required inputs and coordination from the operators. Battle damage assessment is composed of physical damage assessment, functional damage assessment, and target system assessment. Also called BDA. (JP 1-02) In Marine Corps usage, the timely and accurate estimate of the damage resulting from the application of military force. BDA estimates physical damage to a particular target, functional damage to that target, and the capability of the entire target system to continue its operations. (MCRP 5-12C)

battlespace—All aspects of air, surface, subsurface, land, space, and electromagnetic spectrum which encompass the area of influence and area of interest. (MCRP 5-12C)

battlespace dominance—The degree of control over the dimensions of the battlespace which enhances friendly freedom of action and denies enemy freedom of action. It permits force sustainment and application of power projection to accomplish the full range of potential operational and tactical missions. It includes all actions conducted against enemy capabilities to influence future operations. (MCRP 5-12C)

biographical intelligence—That component of intelligence which deals with individual foreign personalities of actual or potential importance. (JP 1-02)

black list—An official counterintelligence listing of actual or potential enemy collaborators, sympathizers, intelligence suspects, and other persons whose presence menaces the security of friendly forces. (JP 1-02) Currently known as the DETAIN category of the Personalities Database within DCIIS.

border crosser—An individual, living close to a frontier, who normally has to cross the frontier frequently for legitimate purposes. (JP 1-02)

bug—**1.** A concealed microphone or listening device or other audiosurveillance device. **2.** To install means for audiosurveillance. (JP 1-02)

bugged—Room or object which contains a concealed listening device. (JP 1-02)

case—**1.** An intelligence operation in its entirety. **2.** Record of the development of an intelligence operation, including personnel, modus operandi, and objectives. (JP 1-02)

cell—Small group of individuals who work together for clandestine or subversive purposes. (JP 1-02)

centers of gravity—Those characteristics, capabilities, or localities from which a military force derives its freedom of action, physical strength, or will to fight. Also called COGs. (JP 1-02).

centralized control—In military operations, a mode of battlespace management in which one echelon of command exercises total authority and direction of all aspects of one or more warfighting functions. It is a method of control where detailed orders are issued and total unity of action is the overriding consideration. (MCRP 5-12C)

clandestine operation—An operation sponsored or conducted by governmental departments or agencies in such a way as to assure secrecy or concealment. A clandestine operation differs from a covert operation in that emphasis is placed on concealment of the operation rather than on concealment of identity of sponsor. In special operations, an activity may be both covert and clandestine and may focus equally on operational considerations and intelligence-related activities. See also covert operation; overt operation. (JP 1-02)

classification—The determination that official information requires, in the interests of national security, a specific degree of protection against unauthorized disclosure, coupled with a designation signifying that such a determination has been made. (JP 1-02)

classified information—Official information which has been determined to require, in the interests of national security, protection against unauthorized disclosure and which has been so designated. (JP 1-02)

code word—**1.** A word that has been assigned a classification and a classified meaning to safeguard intentions and information regarding a classified plan or operation. **2.** A cryptonym used to identify sensitive intelligence data. (JP 1-02)

collection—Acquisition of information and the provision of this information to processing and/or production elements. (JP 1-02) In Marine Corps usage, the gathering of intelligence data and information to satisfy the identified requirements. (MCRP 5-12C)

collection (acquisition)—The obtaining of information in any manner, including direct observation, liaison with official agencies, or solicitation from official, unofficial, or public sources. (JP 1-02)

collection agency—Any individual, organization, or unit that has access to sources of information and the capability of collecting information from them. (JP 1-02)

collection asset—A collection system, platform, or capability that is supporting, assigned, or attached to a particular commander. (JP 1-02)

collection management—The process of converting intelligence requirements into collection requirements, establishing priorities, tasking or coordinating with appropriate collection sources or agencies, monitoring results, and retasking, as required. (JP 1-02)

collection management authority—Constitutes the authority to establish, prioritize and validate theater collection requirements, establish sensor tasking guidance and develop theater collection plans. (JP 1-02)

collection manager—An individual with responsibility for the timely and efficient tasking of organic collection resources and the development of requirements for theater and national assets that could satisfy specific information needs in support of the mission. Also called CM. (JP 1-02)

collection plan—A plan for collecting information from all available sources to meet intelligence requirements and for transforming those requirements into orders and requests to appropriate agencies. (JP 1-02)

collection requirement—An established intelligence need considered in the allocation of intelligence resources to fulfill the essential elements of information and other intelligence needs of a commander. (JP 1-02)

collection requirements management—The authoritative development and control of collection, processing, exploitation, and/or reporting requirements that normally result in either the direct tasking of assets over which the collection manager has authority, or the generation of tasking requests to collection management authorities at a higher, lower, or lateral echelon to accomplish the collection mission. Also called CRM. (JP 1-02)

combat data—Data derived from reporting by operational units. (MCRP 5-12C)

combat service support element—The core element of Marine air-ground task force that is task-organized to provide the combat service support necessary to accomplish the Marine air-ground task force mission. The combat service support element varies in size from a small detachment to one or more force service support groups. It provides supply, maintenance, transportation, general engineering, health services, and a variety of other services to the Marine air-ground task force. It may also contain other Service or foreign military forces assigned or attached to the MAGTF. The combat service support element itself is not a formal command. Also called CSSE. (Approved for inclusion in next version of MCRP 5-12C)

combat surveillance—A continuous, all-weather, day-and-night, systematic watch over the battle area to provide timely information for tactical combat operations. (JP 1-02)

combatant command—A unified or specified command with a broad continuing mission under a single commander established and so designated by the President, through the Secretary of Defense and with the advice and assistance of the Chairman of the Joint Chiefs of Staff. Combatant commands typically have geographic or functional responsibilities. (JP 1-02)

command and control—The exercise of authority and direction by a properly designated commander over command and control assigned and attached forces in the accomplishment of the mission. Command and control functions are performed through an arrangement of personnel, equipment, communications, facilities, and procedures employed by a commander in planning, directing, coordinating, and controlling forces and operations in the accomplishment of the mission. Also called C2. (JP 1-02) Also in Marine Corps usage, the means by which a commander recognizes what needs to be done and sees to it that appropriate actions are taken. (MCRP 5-12C)

command element—The core element of a Marine air-ground task force that is the headquarters. The command element is composed of the commander, general or executive and special staff sections, headquarters section, and requisite communications support, intelligence and reconnaissance forces, necessary to accomplish the MAGTF's mission. The command element provides command and control, intelligence, and other support essential for effective planning and execution of operations by the other elements of the Marine air-ground task force. The command element varies in size and composition and may contain other Service or foreign military forces assigned or attached to the MAGTF. Also called CE. (Approved for inclusion in next version of MCRP 5-12C)

commander's critical information requirements—Information regarding the enemy and friendly activities and the environment identified by the commander as critical to maintaining situational awareness, planning future activities, and facilitating timely decisionmaking. Also called CCIR. NOTE: CCIRs are normally divided into three primary subcategories: priority intelligence requirements; friendly force information requirements; and essential elements of friendly information. (MCRP 5-12C)

commander's intent—commander's clear, concise articulation of the purpose(s) behind one or more tasks assigned to a subordinate. It is one of two parts of every mission statement which guides the exercise of initiative in the absence of instructions. (MCRP 5-12C)

communications intelligence—Technical and intelligence information derived from foreign communications by other than the intended recipients. Also called COMINT. (JP 1-02)

communications intelligence data base—The aggregate of technical and intelligence information derived from the interception and analysis of foreign communications (excluding press, propaganda, and public broadcast) used in the direction and redirection of communications intelligence intercept, analysis, and reporting activities. (JP 1-02)

communications security—The protection resulting from all measures designed to deny unauthorized persons information of value which might be derived from the possession and study of telecommunications, or to mislead unauthorized persons in their interpretation of the results of such possession and study. Also called COMSEC. Communications security includes:

1. cryptosecurity - The component of communications security that results from the provision of technically sound crypto systems and their proper use.

2. transmission security -The component of communications security that results from all measures designed to protect transmissions from interception and exploitation by means other than cryptanalysis.

3. emission security -The component of communications security that results from all measures taken to deny unauthorized persons information of value that might be derived from intercept and analysis of compromising emanations from crypto-equipment and telecommunications systems.

4. physical security-The component of communications security that results from all physical measures necessary to safeguard classified equipment, material, and documents from access thereto or observation thereof by unauthorized persons. (JP 1-02)

compartmentation—Establishment and management of an organization so that information about the personnel, internal organization, or activities of one component is made available to any other component only to the extent required for the performance of assigned duties. (JP 1-02)

complaint-type investigation—A counterintelligence investigation in which sabotage, espionage, treason, sedition, subversive activity, or disaffection is suspected. (JP 1-02)

compromise—The known or suspected exposure of clandestine personnel, installations, or other assets or of classified information or material, to an unauthorized person. (JP 1-02)

compromised—A term applied to classified matter, knowledge of which has, in whole or in part, passed to an unauthorized person or persons, or which has been subject to risk of such passing. (JP 1-02)

confidential—National security information or material which requires protection and the unauthorized disclosure of which could reasonably be expected to cause damage to the national security. (JP 1-02)

confirmation of information (intelligence)—An information item is said to be confirmed when it is reported for the second time, preferably by another independent source whose reliability is considered when confirming information. (JP 1-02)

confusion agent—An individual who is dispatched by the sponsor for the primary purpose of confounding the intelligence or counterintelligence apparatus of another country rather than for the purpose of collecting and transmitting information. (JP 1-02)

contingency—An emergency involving military forces caused by natural disasters, terrorists, subversives, or by required military operations. Due to the uncertainty of the situation, contingencies require plans, rapid response, and special procedures to ensure the safety and readiness of personnel, installations, and equipment. (JP 1-02)

control—**1.** Authority which may be less than full command exercised by a commander over part of the activities of subordinate or other organizations. **2.** In mapping, charting, and photogrammetry, a collective term for a system of marks or objects on the earth or on a map or a photograph, whose positions or elevations, or both, have been or will be determined. **3.** Physical or psychological pressures exerted with the intent to assure that an agent or group will respond as directed. **4.** An indicator governing the distribution and use of documents, information, or material. Such indicators are the subject of intelligence community agreement and are specifically defined in appropriate regulations. (JP 1-02)

controlled information—Information conveyed to an adversary in a deception operation to evoke desired appreciations. (JP 1-02)

coordinating authority—A commander or individual assigned responsibility for coordinating specific functions or activities involving forces of two or more Military Departments or two or more forces of the same Service. The commander or individual has the authority to require consultation between the agencies involved, but does not have the authority to compel agreement. In the event that essential agreement cannot be obtained, the matter shall be referred to the appointing authority. Coordinating authority is a consultation relationship, not an authority through which command may be exercised. Coordinating authority is more applicable to planning and similar activities than to operations. (JP 1-02)

coordination—The action necessary to ensure adequately integrated relationships between separate organizations located in the same area. Coordination may include such matters as fire support, emergency defense measures, area intelligence, and other situations in which coordination is considered necessary. (MCRP 5-12C)

counterdeception—Efforts to negate, neutralize, diminish the effects of, or gain advantage from, a foreign deception operation. Counterdeception does not include the intelligence function of identifying foreign deception operations. (JP 1-02)

counterespionage—That aspect of counterintelligence designed to detect, destroy, neutralize, exploit, or prevent espionage activities through identification, penetration, manipulation, deception, and repression of individuals, groups, or organizations conducting or suspected of conducting espionage activities. (JP 1-02)

counterguerrilla warfare—Operations and activities conducted by armed forces, paramilitary forces, or nonmilitary agencies against guerrillas. (JP 1-02)

counterinsurgency—Those military, paramilitary, political, economic, psychological, and civic actions taken by a government to defeat insurgency. (JP 1-02)

counterintelligence—**1.** Information gathered and activities conducted to protect against espionage, other intelligence activities, sabotage, or assassinations conducted by or on behalf of foreign governments or elements thereof, foreign organizations, or foreign persons, or international terrorist activities. Also called CI. See also counterespionage; security. (JP 1-02) **2.** Within the Marine Corps, counterintelligence (CI) constitutes active and passive measures intended to deny a threat force valuable information about the friendly situation, to detect and neutralize hostile intelligence collection, and to deceive the enemy as to friendly capabilities and intentions. (MCRP 5-12C)

counterintelligence activities—The four functions of counterintelligence: operations; investigations; collection and reporting; and analysis, production, and dissemination. See also counterintelligence. (JP 1-02)

counterintelligence collection—The systematic acquisition of information (through investigations, operations, or liaison) concerning espionage, sabotage, terrorism, other intelligence activities or assassinations conducted by or on behalf of foreign governments or elements thereof, foreign organizations, or foreign persons which are directed against or threaten Department of Defense interests. Includes liaison and CFSO. (JP 1-02)

counterintelligence force protection source operations—Collection activities conducted by CI personnel to provide force protection support. These operations respond to local command requirements for force protection and do not fall within the purview of DCID 5/1. Also called CFSO. (MCRP 5-12C)

counterintelligence investigations—Counterintelligence investigations establish the elements of proof for prosecution or administrative action. Counterintelligence investigations can provide a basis for or be developed from conducting counterintelligence operations. Counterintelligence investigations are conducted against individuals or groups suspected of committing acts of espionage, sabotage, sedition, subversion, terrorism, and other major security violations as well as failure to follow Defense agency and military Service directives governing reporting of contacts with foreign citizens and "out-of-channel" requests for defense information. Counterintelligence investigations provide military commanders and policymakers with information used to eliminate security vulnerabilities and otherwise to improve the security posture of threatened interests. See also counterintelligence. (JP 2-01.2)

counterintelligence production—The process of analyzing all-source information concerning espionage, or other multidiscipline intelligence collection threats, sabotage, terrorism, and other related threats to US military commanders, the Department of Defense, and the US Intelligence Community and developing it into a final product which is disseminated. Counterintelligence production is used in formulating security policy, plans, and operations. See also counterintelligence. (JP 1-02)

countermeasures—That form of military science that, by the employment of devices and/or techniques, has as its objective the impairment of the operational effectiveness of enemy activity. (JP 1-02)

counterreconnaissance—All measures taken to prevent hostile observation of a force, area, or place. (JP 1-02)

countersabotage—That aspect of counterintelligence designed to detect, destroy, neutralize, or prevent sabotage activities through identification, penetration, manipulation, deception, and repression of individuals, groups, or organizations conducting or suspected of conducting sabotage activities. (JP 1-02)

countersign—A secret challenge and its reply. (JP 1-02)

countersubversion—That aspect of counterintelligence designed to detect, destroy, neutralize, or prevent subversive activities through the identification, exploitation, penetration, manipulation, deception, and repression of individuals, groups, or organizations conducting or suspected of conducting subversive activities. (JP 1-02)

counterterrorism—Offensive measures taken to prevent, deter, and respond to terrorism. Also called CT. (JP 1-02)

Country Team—The senior, in-country, United States coordinating and supervising body, headed by the Chief of the United States diplomatic mission, and composed of the senior member of each represented United States department or agency, as desired by the Chief of the US diplomatic mission. (JP 1-02)

cover—**1.** The action by land, air, or sea forces to protect by offense, defense, or threat of either or both. **2.** Those measures necessary to give protection to a person, plan, operation, formation or installation from the enemy intelligence effort and leakage of information. **3.** The act of maintaining a continuous receiver watch with transmitter calibrated and available, but not necessarily available for immediate use. **4.** Shelter or protection, either natural or artificial. **5.** Photographs or other recorded images which show a particular area of ground. **6.** A code meaning, "Keep fighters between force/base and contact designated at distance stated from force/base" (e.g., "cover bogey twenty-seven to thirty miles"). (JP 1-02)

cover (military)—Actions to conceal actual friendly intentions, capabilities, operations, and other activities by providing a plausible, yet erroneous, explanation of the observable. (JP 1-02)

covert operations—An operation that is so planned and executed as to conceal the identity of or permit plausible denial by the sponsor. A covert operation differs from a clandestine operation in that emphasis is placed on concealment of identity of sponsor rather than on concealment of the operation. See also clandestine operation; overt operation. (JP 1-02)

critical information—Specific facts about friendly intentions, capabilities, and activities vitally needed by adversaries for them to plan and act effectively so as to guarantee failure or unacceptable consequences for friendly mission accomplishment. (JP 1-02)

critical vulnerability—An aspect of a center of gravity that if exploited will do the most significant damage to an adversary's ability to resist. A vulnerability cannot be critical unless it undermines a key strength. Also called CV. (MCRP 5-12C)

cultivation—A deliberate and calculated association with a person for the purpose of recruitment, obtaining information, or gaining control for these or other purposes. (JP 1-02)

current intelligence—Intelligence of all types and forms of immediate interest which is usually disseminated without the delays necessary to complete evaluation or interpretation. (JP 1-02)

damage assessment—(1) The determination of the effect of attacks on targets. (2) A determination of the effect of a compromise of classified information on national security. (JP 1-02)

data base—Information that is normally structured and indexed for user access and review. Data bases may exist in the form of physical files (folders, documents, etc.) or formatted automated data processing system data files. (JP 1-02)

deception—Those measures designed to mislead the enemy by manipulation, distortion, or falsification of evidence to induce him to react in a manner prejudicial to his interests. (JP 1-02)

decentralized control—In military operations, a mode of battlespace management in which a command echelon may delegate some or all authority and direction for warfighting functions to subordinates. It requires careful and clear articulation of mission, intent, and main effort to unify efforts of subordinate leaders. (MCRP 5-12C)

declassification—The determination that in the interests of national security, classified information no longer requires any degree of protection against unauthorized disclosure, coupled with removal or cancellation of the classification designation. (JP 1-02)

departmental intelligence—Intelligence that any department or agency of the Federal Government requires to execute its own mission. (JP 1-02)

Department of Defense Intelligence Information System—The aggregation of DOD personnel, procedures, equipment, computer programs, and supporting communications that support the timely and comprehensive preparation and presentation of intelligence and intelligence information to military commanders and national-level decisionmakers. Also called DODIIS. (JP 1-02)

descriptive intelligence—Class of intelligence which describes existing and previously existing conditions with the intent to promote situational awareness. Descriptive intelligence has two components: basic intelligence, which is general background knowledge about established and relatively constant conditions; and current intelligence, which is concerned with describing the existing situation. (MCRP 5-12C)

detachment—**1.** A part of a unit separated from its main organization for duty elsewhere. **2.** A temporary military or naval unit formed from other units or parts of units. (JP 1-02)

detection—**1.** In tactical operations, the perception of an object of possible military interest but unconfirmed by recognition. **2.** In surveillance, the determination and transmission by a surveillance system that an event has occurred. **3.** In arms control, the first step in the process of ascertaining the occurrence of a violation of an arms-control agreement. (JP 1-02)

disaffected person—A person who is alienated or estranged from those in authority or lacks loyalty to the government; a state of mind. (JP 1-02)

dissemination—Conveyance of intelligence to users in a suitable form. (JP 1-02)

dissemination management—Involves establishing dissemination priorities, selection of dissemination means, and monitoring the flow of intelligence throughout the command. The objective of dissemination management is to deliver the required intelligence to the appropriate user in proper form at the right time while ensuring that individual consumers and the dissemination system are not overloaded attempting to move unneeded or irrelevant information. Dissemination management also provides for use of security controls which do not impede the timely delivery or subsequent use of intelligence while protecting intelligence sources and methods. (MCRP 5-12C)

domestic intelligence—Intelligence relating to activities or conditions within the United States that threaten internal security and that might require the employment of troops; and intelligence relating to activities of individuals or agencies potentially or actually dangerous to the security of the Department of Defense. (JP 1-02)

double agent—Agent in contact with two opposing intelligence services, only one of which is aware of the double contact or quasi-intelligence services. (JP 1-02)

espionage—The act of obtaining, delivering, transmitting, communicating, or receiving information about the national defense with an intent, or reason to believe, that the information may be used to the injury of the United States or to the advantage of any foreign nation. (JP 1-02)

espionage against the United States—Overt, covert, or clandestine activity designed to obtain information relating to the national defense with intent or reason to believe that it will be used to the injury of the United States or to the advantage of a foreign nation. For espionage crimes see Chapter 37 of Title 18, United States Code. (JP 1-02)

essential elements of friendly information—Key questions likely to be asked by adversary officials and intelligence systems about specific friendly intentions, capabilities, and activities so they can obtain answers critical to their operational effectiveness. Also called EEFI. (JP 1-02) Specific facts about friendly intentions, capabilities, and activities needed by adversaries to plan and execute effective operations against our forces. (MCRP 5-12C)

estimate—**1.** An analysis of a foreign situation, development, or trend that identifies its major elements, interprets the significance, and appraises the future possibilities and the prospective results of the various actions that might be taken. **2.** An appraisal of the capabilities, vulnerabilities, and potential courses of action of a foreign nation or combination of nations in consequence of a specific national plan, policy, decision, or contemplated course of action. **3.** An analysis of an actual or contemplated clandestine operation in relation to the situation in which it is or would be conducted in order to identify and appraise such factors as available and needed assets and potential obstacles, accomplishments, and consequences. (Excerpt from JP 1-02)

estimative intelligence—Class of intelligence which attempts to anticipate future possibilities and probabilities based on an analysis of descriptive intelligence in the context of planned friendly and assessed enemy operations. (MCRP 5-12C)

evaluation—In intelligence usage, appraisal of an item of information in terms of credibility, reliability, pertinence, and accuracy. Appraisal is accomplished at several stages within the intelligence cycle with progressively different contexts. Initial evaluations, made by case officers and report officers, are focused upon the reliability of the source and the accuracy of the information as judged by data available at or close to their operational levels. Later evaluations by intelligence analysts are primarily concerned with verifying accuracy of information and may, in effect, convert information into intelligence. Appraisal or evaluation of items of information or intelligence is indicated by a standard letter-number system. The evaluation of the reliability of sources is designated by a letter from A through F, and the accuracy of the information is designated by numeral 1 through 6. These are two entirely independent appraisals, and these separate appraisals are indicated in accordance with the system indicated below. Thus, information adjudged to be "probably true" received from an "usually reliable source" is designated "B-2" or "B2," while information of which the "truth cannot be judged" received from "usually reliable source" is designated "B-6" or "B6."

Reliability of Source

A - Completely reliable

B - Usually reliable

C - Fairly reliable

D - Not usually reliable

E - Unreliable

F - Reliability cannot be judged

Accuracy of Information

1 - Confirmed by other sources

2 - Probably true

3 - Possibly true

4 - Doubtful

5 - Improbable

6 - Truth cannot be judged (JP 1-02)

evasion and escape intelligence—Processed information prepared to assist personnel to escape if captured by the enemy or to evade capture if lost in enemy-dominated territory. (JP 1-02)

fabricator— Individuals or groups who, without genuine resources, invent information or inflate or embroider over news for personal gain or for political purposes. (JP 1-02)

force protection—Security program designed to protect Service members, civilian employees, family members, facilities, and equipment, in all locations and situations, accomplished through planned and integrated application of combatting terrorism, physical security, operations security, personal protective services, and supported by intelligence, CI, and other security programs. (JP 1-02)

foreign intelligence—Information relating to capabilities, intentions, and activities of foreign powers, organizations, or persons, but not including counterintelligence (except for information on international terrorist activities). (JP 1-02)

friendly force information requirements—Information the commander needs about friendly forces in order to develop plans and make effective decisions. Depending upon the circumstances, information on unit location, composition, readiness, personnel status, and logistics status could become a friendly force information requirement. Also called FFIR. (MCRP 5-12C)

fusion—In intelligence usage, the process of examining all sources of intelligence and information to derive a complete assessment of activity. (JP 1-02)

global sourcing—A process of force provision or augmentation whereby resources may be drawn from any location/command worldwide. (MCRP 5-12C)

ground combat element—The core element of a Marine air-ground task force that is task-organized to conduct ground operations. It is usually constructed around an infantry organization but can vary in size from a small ground unit of any type, to one or more Marine divisions that can be independently maneuvered under the direction of the MAGTF commander. It includes appropriate ground combat and combat support forces and may contain other Service or foreign military forces assigned or attached to the Marine air-ground task force. The ground combat element itself is not a formal command. Also called GCE. (Approved for inclusion in next version of MCRP 5-12C)

high-payoff target—A target whose loss to the enemy will significantly contribute to the success of the friendly course of action. High-payoff targets are those high-value targets, identified through wargaming, which must be acquired and successfully attacked for the success of the friendly commander's mission. Also called HPT. (JP 1-02)

high-value target—A target the enemy commander requires for the successful completion of the mission. The loss of high-value targets would be expected to seriously degrade important enemy functions throughout the friendly commander's area of interest. Also called HVT. (JP 1-02)

host country—A nation in which representatives or organizations of another state are present because of government invitation and/or international agreement. (JP 1-02)

host nation—A nation which receives the forces and/or supplies of allied nations and/or NATO organizations to be located on, to operate in, or to transit through its territory. (JP 1-02)

hostage—A person held as a pledge that certain terms or agreements will be kept. (The taking of hostages is forbidden under the Geneva Conventions, 1949.) (JP 1-02)

human intelligence—A category of intelligence derived from information collected and provided by human sources. Also called HUMINT. (JP 1-02) In Marine Corps usage, HUMINT operations cover a wide range of activities encompassing reconnaissance patrols, aircrew reports and debriefs, debriefing of refugees, interrogations of prisoners of war, and the conduct of CI force protection source operations. (MCRP 5-12C)

human resources intelligence—The intelligence information derived from the intelligence collection discipline that uses human beings as both sources and collectors, and where the human being is the primary collection instrument. Also called HUMINT. (JP 1-02)

imagery—Collectively, the representations of objects reproduced electronically or by optical means on film, electronic display devices, or other media. (JP 1-02)

imagery exploitation—The cycle of processing and printing imagery to the positive or negative state, assembly into imagery packs, identification, interpretation, mensuration, information extraction, the preparation of reports, and the dissemination of information. (JP 1-02)

imagery intelligence—Intelligence derived from the exploitation of collection by visual photography, infrared sensors, lasers, electro-optics and radar sensors such as synthetic aperture radar wherein images of objects are reproduced optically or electronically on film, electronic display devices or other media. Also called IMINT. (JP 1-02)

imagery interpretation—**1.** The process of location, recognition, identification, and description of objects, activities, and terrain represented on imagery. **2.** The extraction of information from photographs or other recorded images. (JP 1-02)

imitative deception—The introduction of electromagnetic energy into enemy systems that imitates enemy emissions. (JP 1-02)

indications and warning—Those intelligence activities intended to detect and report time-sensitive intelligence information on foreign developments that could involve a threat to the United States or allied/coalition military, political, or economic interests or to U.S. citizens abroad. It includes forewarning of enemy actions or intentions; the imminence of hostilities; insurgency; nuclear/non-nuclear attack on the United States, its overseas forces, or allied/coalition nations; hostile reactions to United States reconnaissance activities; terrorists' attacks; and other similar events. Also called I&W. (JP 1-02)

indications (intelligence)—Information in various degrees of evaluation, all of which bears on the intention of a potential enemy to adopt or reject a course of action. (JP 1-02)

indicator—In intelligence usage, an item of information which reflects the intention or capability of a potential enemy to adopt or reject a course of action. (JP 1-02)

infiltration—**1.** The movement through or into an area or territory occupied by either friendly or enemy troops or organizations. The movement is made, either by small groups or by individuals, at extended or irregular intervals. When used in connection with the enemy, it infers that contact is avoided. **2.** In intelligence usage, placing an agent or other person in a target area in hostile territory. Usually involves crossing a frontier or other guarded line. Methods of infiltration are: black (clandestine); gray (through legal crossing point but under false documentation); white (legal). (JP 1-02)

informant—(1) A person who, wittingly or unwittingly, provides information to an agent, a clandestine service, or the police. (2) In reporting, a person who has provided specific information and is cited as a source. (JP 1-02)

information—**1.** Facts, data, or instructions in any medium or form. **2.** The meaning that a human assigns to data by means of the known conventions used in their representation. (JP 1-02)

information exchange requirement—The requirement for information to be passed between and among forces, organizations, or administrative structures concerning ongoing activities. Information exchange requirements identify who exchanges what information with whom, as well as why the information is necessary and how that information will be used. The quality (i.e., frequency, timeliness, security) and quantity (i.e., volume, speed, and type of information such as data, voice, and video) are attributes of the information exchange included in the information exchange requirement. Also called IER. (MCRP 5-12C)

informer—Person who intentionally discloses to police or to a security service information about persons or activities considered suspect, usually for a financial reward. (JP 1-02)

infrared imagery—That imagery produced as a result of sensing electromagnetic radiations emitted or reflected from a given target surface in the infrared position of the electromagnetic spectrum (approximately 0.72 to 1,000 microns). (JP 1-02)

insurgency—An organized movement aimed at the overthrow of a constituted government through use of subversion and armed conflict. (JP 1-02)

integration—**1.** A stage in the intelligence cycle in which a pattern is formed through the selection and combination of evaluated information. **2.** In photography, a process by which the average radar picture seen on several scans of the time base may be obtained on a print, or the process by which several photographic images are combined into a single image. (JP 1-02)

intelligence—**1.** The product resulting from the collection, processing, integration, analysis, evaluation, and interpretation of available information concerning foreign countries or areas. **2.** Information and knowledge about an adversary obtained through observation, investigation, analysis, or understanding. (JP 1-02) Also in Marine Corps usage, intelligence is knowledge about the enemy or the surrounding environment needed to support decisionmaking. This knowledge is the result of the collection, processing, exploitation, evaluation, integration, analysis, and interpretation of available information about the battlespace and threat. (MCRP 5-12C)

intelligence annex—A supporting document of an operation plan or order that provides detailed information on the enemy situation, assignment of intelligence tasks, and intelligence administrative procedures. (JP 1-02)

intelligence contingency funds—Appropriated funds to be used for intelligence activities when the use of other funds is not applicable or would either jeopardize or impede the mission of the intelligence unit. (JP 1-02)

intelligence cycle—The steps by which information is converted into intelligence and made available to users. (Excerpt from JP 1-02)

intelligence data—Data derived from assets primarily dedicated to intelligence collection such as imagery systems, electronic intercept equipment, human intelligence sources, etc. (MCRP 5-12C)

intelligence data base—The sum of holdings of intelligence data and finished intelligence products at a given organization. (JP 1-02)

intelligence discipline—A well-defined area of intelligence collection, processing, exploitation, and reporting using a specific category of technical or human resources. There are five major disciplines: human intelligence, imagery intelligence, measurement and signature intelligence, signals intelligence (communications intelligence, electronic intelligence, and foreign instrumentation signals intelligence), and open-source intelligence. (JP 1-02)

intelligence estimate—The appraisal, expressed in writing or orally, of available intelligence relating to a specific situation or condition with a view to determining the courses of action open to the enemy or potential enemy and the order of probability of their adoption. (JP 1-02)

intelligence operations—The variety of intelligence tasks that are carried out by various intelligence organizations and activities. (Excerpt from JP 1-02)

intelligence preparation of the battlespace—An analytical methodology employed to reduce uncertainties concerning the enemy, environment, and terrain for all types of operations. Intelligence preparation of the battlespace builds an extensive data base for each potential area in which a unit may be required to operate. The data base is then analyzed in detail to determine the impact of the enemy, environment, and terrain on operations and presents it in graphic form. Intelligence preparation of the battlespace is a continuing process. Also called IPB. (JP 1-02) In Marine Corps usage, the systematic, continuous process of analyzing the threat and environment in a specific geographic area. (MCRP 5-12C)

intelligence-related activities—**1.** Those activities outside the consolidated defense intelligence program which: a. Respond to operational commanders' tasking for time-sensitive information on foreign entities; b. Respond to national intelligence community tasking of systems whose primary mission is support to operating forces; c. Train personnel for intelligence duties; d. Provide an intelligence reserve; or e. Are devoted to research and development of intelligence or related capabilities. **2.** Specifically excluded are programs which are so closely integrated with a weapon system that their primary function is to provide immediate-use targeting data. (JP 1-02)

intelligence report—A specific report of information, usually on a single item, made at any level of command in tactical operations and disseminated as rapidly as possible in keeping with the timeliness of the information. Also called INTREP. (JP 1-02)

intelligence reporting—The preparation and conveyance of information by any means. More commonly, the term is restricted to reports as they are prepared by the collector and as they are transmitted by the collector to the latter's headquarters and by this component of the intelligence structure to one or more intelligence-producing components. Thus, even in this limited sense, reporting embraces both collection and dissemination. The term is applied to normal and specialist intelligence reports. (JP 1-02)

intelligence requirement—Any subject, general or specific, upon which there is a need for the collection of information, or the production of intelligence. Also called IR. (JP 1-02) In Marine Corps usage, questions about the enemy and the environment, the answers to which a commander requires to make sound decisions. (MCRP 5-12C)

intelligence system—Any formal or informal system to manage data gathering, to obtain and process the data, to interpret the data, and to provide reasoned judgments to decisionmakers as a basis for action. The term is not limited to intelligence organizations or services but includes any system, in all its parts, that accomplishes the listed tasks. (JP 1-02)

internal security—The state of law and order prevailing within a nation. (JP 1-02)

intelligence summary—A specific report providing a summary of items of intelligence at frequent intervals. (JP 1-02)

interoperability—**1.** The ability of systems, units or forces to provide services to and accept services from other systems, units, or forces and to use the services so exchanged to enable them to operate effectively together. (DOD) **2.** The condition achieved among communications-electronics systems or items of communications-electronics equipment when information or services can be exchanged directly and satisfactorily between them and/or their users. The degree of interoperability should be defined when referring to specific cases. (JP 1-02)

interpretation—A stage in the intelligence cycle in which the significance of information is judged in relation to the current body of knowledge. (JP 1-02)

interrogation (intelligence)—Systematic effort to procure information by direct questioning of a person under the control of the questioner. (JP 1-02)

interview (intelligence)—To gather information from a person who is aware that information is being given although there is ignorance of the true connection and purposes of the interviewer. Generally overt unless the collector is other than purported to be. (JP 1-02)

investigation—A duly authorized, systematized, detailed examination or inquiry to uncover facts and determine the truth of a matter. This may include collecting, processing, reporting, storing, recording, analyzing, evaluating, producing and disseminating the authorized information. (JP 1-02)

J-2X—Umbrella organization consisting of the HUMINT Operations Cell and the Task Force Counterintelligence Coordinating Authority. The J-2X is responsible for coordination and deconfliction of all human source related activity. (JP 1-02)

Joint Deployable Intelligence Support System—A transportable workstation and communications suite that electronically extends a joint intelligence center to a joint task force or other tactical user. Also called JDISS. (JP 1-02)

joint document exploitation center—Physical location for deriving intelligence information from captured enemy documents. It is normally subordinate to the joint force/J-2. Also called JDEC. (JP 1-02)

joint force—A general term applied to a force composed of significant elements, assigned or attached, of two or more Military Departments, operating under a single joint force commander. (JP 1-02)

joint force commander—A general term applied to a combatant commander, subunified commander, or joint task force commander authorized to exercise combatant command (command authority) or operational control over a joint force. Also called JFC. (JP 1-02)

joint intelligence—Intelligence produced by elements of more than one Service of the same nation. (JP 1-02)

joint intelligence architecture—A dynamic, flexible structure that consists of the National Military Joint Intelligence Center, the theater joint intelligence centers, and subordinate joint force joint intelligence support elements. This architecture encompasses automated data processing equipment capabilities, communications and information requirements, and responsibilities to provide national, theater, and tactical commanders with the full range of intelligence required for planning and conducting operations. (JP 1-02)

joint intelligence center—The intelligence center of the joint force headquarters. The joint intelligence center is responsible for providing and producing the intelligence required to support the joint force commander and staff, components, task forces and elements, and the national intelligence community. Also called JIC. (JP 1-02)

joint intelligence element—A subordinate joint force forms a joint intelligence support element as the focus for intelligence support for joint operations, providing the joint force commander, joint staff, and components with the complete air, space, ground, and maritime adversary situation. Also called JISE. (JP 1-02)

joint interrogation and debriefing center—Physical location for the exploitation of intelligence information from enemy prisoners of war and other non-prisoner sources. It is normally subordinate to the joint force/J-2. Also called JIDC. (JP 1-02)

joint operational intelligence agency—An intelligence agency in which the efforts of two or more Services are integrated to furnish that operational intelligence essential to the commander of a joint force and to supplement that available to subordinate forces of the command. The agency may or may not be part of such joint force commander's staff. (JP 1-02)

Joint Worldwide Intelligence Communications System—The sensitive compartmented information portion of the Defense Information System Network. It incorporates advanced networking technologies that permit point-to-point or multipoint information exchange involving voice, text, graphics, data, and video teleconferencing. Also called JWICS. (JP 1-02)

law of war—That part of international law that regulates the conduct of armed hostilities. Also called the law of armed conflict. (JP 1-02)

liaison—that contact or intercommunication maintained between elements of military forces or other agencies to ensure mutual understanding and unity of purpose and action. (JP 1-02)

main effort—The designated subordinate unit whose mission at a given point in time is most critical to overall mission success. It is usually weighted with the preponderance of combat power and is directed against a center of gravity through a critical vulnerability. (MCRP 5-12C)

maneuver warfare—A warfighting philosophy that seeks to shatter the enemy's cohesion through a variety of rapid, focused, and unexpected actions which create a turbulent and rapidly deteriorating situation with which the enemy cannot cope. (MCRP 5-12C)

Marine air-ground task force—The Marine Corps principal organization for all missions across the range of military operations, composed of forces task-organized under a single commander capable of responding rapidly to a contingency anywhere in the world. The types of forces in the MAGTF are functionally grouped into four core elements: a command element, an aviation combat element, a ground combat element, and a combat service support element. The four core elements are categories of forces, not formal commands. The basic structure of the Marine air-ground task force never varies, though the number, size, and type of Marine Corps units comprising each of its four elements will always be mission dependent. The flexibility of the organizational structure allows for one or more subordinate MAGTFs, other Service and/or foreign military forces, to be assigned or attached. Also called MAGTF. (Approved for inclusion in next version of MCRP 5-12C)

Marine Corps Planning Process—A six-step methodology which helps organize the thought processes of the commander and staff throughout the planning and execution of military operations. It focuses on the threat and is based on the Marine Corps philosophy of maneuver warfare. It capitalizes on the principle of unity of command and supports the establishment and maintenance of tempo. The six steps consist of mission analysis, course of action development, course of action analysis, comparison/decision, orders development, and transition. Also called MCPP. NOTE: Tenets of the MCPP include top down planning, single battle concept, and integrated planning. (MCRP 5-12C)

Marine expeditionary force—The largest Marine air-ground task force and the Marine Corps principal warfighting organization, particularly for larger crises or contingencies. It is task-organized around a permanent command element and normally contains one or more Marine divisions, Marine aircraft wings, and Marine force service support groups. The Marine expeditionary force is capable of missions across the range of military operations, including amphibious assault and sustained operations ashore in any environment. It can operate from a sea base, a land base, or both. It may also contain other Service or foreign military forces assigned or attached to the MAGTF. Also called MEF. (Approved for inclusion in next version of MCRP 5-12C)

Marine expeditionary force (Forward)—A designated lead echelon of a Marine expeditionary force, task-organized to meet the requirements of a specific situation. A Marine expeditionary force (Forward) varies in size and composition, and may be commanded by the Marine expeditionary force commander personally or by another designated commander. It may be tasked with preparing for the subsequent arrival of the rest of the MEF/joint/combined forces, and/or the conduct of other specified tasks, at the discretion of the MEF commander. A Marine expeditionary force (Forward) may also be a stand-alone MAGTF, task-organized for a mission in which a MEF is not required. It may also contain other Service or foreign military forces assigned or attached to the Marine air-ground task force. Also called MEF (Fwd). (Approved for inclusion in next version of MCRP 5-12C)

Marine expeditionary unit—A Marine air-ground task force that is constructed around an infantry battalion reinforced, a helicopter squadron reinforced, and a task-organized combat service support element. It normally fulfills Marine Corps forward sea-based deployment requirements. The Marine expeditionary unit provides an immediate reaction capability for crisis response and is capable of limited combat operations. It may contain other Service or foreign military forces assigned or attached. Also called MEU. (Approved for inclusion in next version of MCRP 5-12C)

Marine expeditionary unit (special operations capable)—The Marine Corps standard, forward-deployed, sea-based expeditionary organization. The MEU(SOC) is a MEU, augmented with selected personnel and equipment, that is trained and equipped with an enhanced capability to conduct amphibious operations and a variety of specialized missions, of limited scope and duration. These capabilities include specialized demolition, clandestine reconnaissance and surveillance, raids, in-extremis hostage recovery, and enabling operations for follow-on forces. The Marine expeditionary unit (special operations capable) is not a special operations force but, when directed by the National Command Authorities, the combatant commander in chief, and/or other operational commander, may conduct limited special operations in extremis, when other forces are inappropriate or unavailable. It may also contain other Service or foreign military forces assigned or attached to the Marine air-ground task force. Also called MEU (SOC). (Approved for inclusion in next version of MCRP 5-12C)

measurement and signature intelligence—Scientific and technical intelligence obtained by quantitative and qualitative analysis of data (metric, angle, spatial, wavelength, time dependence, modulation, plasma, and hydromagnetic) derived from specific technical sensors for the purpose of identifying any distinctive features associated with the target. The detected feature may be either reflected or emitted. Also called MASINT. (JP 1-02)

military intelligence—Intelligence on any foreign military or military-related situation or activity which is significant to military policy making or the planning and conduct of military operations and activities. (JP 1-02)

Military Intelligence Integrated Data System/ Integrated Data Base—An architecture for improving the manner in which military intelligence is analyzed, stored, and disseminated. The Integrated Data Base (IDB) forms the core automated data base for the Military Intelligence Integrated Data System (MIIDS) program and integrates the data in the installation, order of battle, equipment, and selected electronic warfare and command, control, and communications files. The IDB is the national-level repository for the general military intelligence information available to the entire Department of Defense Intelligence Information System community and maintained by DIA and the commands. The IDB is kept synchronized by system transactions to disseminate updates. Also called MIIDS/IDB. (JP 1-02)

national intelligence—Integrated departmental intelligence that covers the broad aspects of national policy and national security, is of concern to more than one department or agency, and transcends the exclusive competence of a single department or agency. (JP 1-02)

need to know—A criterion used in security procedures which requires the custodians of classified information to establish, prior to disclosure, that the intended recipient must have access to the information to perform his or her official duties. (JP 1-02)

neutralize—As pertains to military operations, to render ineffective or unusable. (JP 1-02)

official information—Information which is owned by, produced for or by, or is subject to the control of the United States Government. (JP 1-02)

open-source intelligence—Information of potential intelligence value that is available to the general public. Also called OSINT. (JP 1-02)

operational control—Transferable command authority that may be exercised by commanders at any echelon at or below the level of combatant command. Operational control is inherent in combatant command (command authority). Operational control may be delegated and is the authority to perform those functions of command over subordinate forces involving organizing and employing commands and forces, assigning tasks, designating objectives, and giving authoritative direction necessary to accomplish the mission. Operational control includes authoritative direction over all aspects of military operations and joint training necessary to accomplish missions assigned to the command. Operational control should be exercised through the commanders of subordinate organizations. Normally this authority is exercised through subordinate joint force commanders and Service and/or functional component commanders. Operational control normally provides full authority to organize commands and forces and to employ those forces as the commander in operational control considers necessary to accomplish assigned missions. Operational control does not, in and of itself, include authoritative direction for logistics or matters of administration, discipline, internal organization, or unit training. Also called OPCON. (JP 1-02)

operation order—A directive issued by a commander to subordinate commanders for the purpose of effecting the coordinated execution of an operation. Also called OPORD. (JP 1-02)

operation plan—Any plan, except for the Single Integrated Operation Plan, for the conduct of military operations. Plans are prepared by combatant commanders in response to requirements established by the Chairman of the Joint Chiefs of Staff and by commanders of subordinate commands in response to requirements tasked by the establishing unified commander. Operation plans are prepared in either a complete format (OPLAN) or as a concept plan (CONPLAN). The CONPLAN can be published with or without a time-phased force and deployment data (TPFDD) file. **a.** OPLAN—An operation plan for the conduct of joint operations that can be used as a basis for development of an operation order (OPORD). An OPLAN identifies the forces and supplies required to execute the CINC's Strategic Concept and a movement schedule of these resources to the theater of operations. The forces and supplies are identified in TPFDD files. OPLANs will include all phases of the tasked operation. The plan is prepared with the appropriate annexes, appendixes, and TPFDD files as described in the Joint Operation Planning and Execution System manuals containing planning policies, procedures, and formats. Also called OPLAN. **b.** CONPLAN—An operation plan in an abbreviated format that would require considerable expansion or alteration to convert it into an OPLAN or OPORD. A CONPLAN contains the CINC's Strategic Concept and those annexes and appendixes deemed necessary by the combatant commander to complete planning. Generally, detailed support requirements are not calculated and TPFDD files are not prepared. Also called CONPLAN. **c.** CONPLAN with TPFDD—A CONPLAN with TPFDD is the same as a CONPLAN except that it requires more detailed planning for phased deployment of forces. (JP 1-02)

operations security—A process of identifying critical information and subsequently analyzing friendly actions attendant to military operations and other activities to:

a. Identify those actions that can be observed by adversary intelligence systems.

b. Determine indicators hostile intelligence systems might obtain that could be interpreted or pieced together to derive critical information in time to be useful to adversaries.

c. Select and execute measures that eliminate or reduce to an acceptable level the vulnerabilities of friendly actions to adversary exploitation. Also called OPSEC. (JP 1-02)

order of battle—The identification, strength, command structure, and disposition of the personnel, units, and equipment of any military force. Also called OOB. (JP 1-02)

overt operation—An operation conducted openly, without concealment. (JP 1-02)

penetration—**1.** In land operations, a form of offensive which seeks to break through the enemy's defense and disrupt the defensive system. (JP 1-02) **2.** The recruitment of agents within, or the infiltration of agents or technical monitoring devices in an organization or group for the purpose of acquiring information or of influencing its activities.

personnel security investigation—An inquiry into the activities of an individual which is designed to develop pertinent information pertaining to trustworthiness and suitability for a position of trust as related to loyalty, character, emotional stability, and reliability. (JP 1-02)

physical security—That part of security concerned with physical measures designed to safeguard personnel, to prevent unauthorized access to equipment, installations, material and documents, and to safeguard them against espionage, sabotage, damage, and theft. (JP 1-02)

positive intelligence—A term of convenience sometimes applied to foreign intelligence to distinguish it from foreign counterintelligence.

principal agent—An agent who, under the direction of an intelligence officer, is responsible for the operational activities of other agents.

priority intelligence requirements—Those intelligence requirements for which a commander has an anticipated and stated priority in his task of planning and decisionmaking. Also called PIR. (JP 1-02) In Marine Corps usage, an intelligence requirement associated with a decision that will critically affect the overall success of the command's mission. (MCRP 5-12C)

production management—Encompasses determining the scope, content, and format of each intelligence product, developing a plan and schedule for the development of each product, assigning priorities among the various production requirements, allocating processing, exploitation, and production resources, and integrating production efforts with intelligence collection and dissemination. (MCRP 5-12C)R

ratline—An organized effort for moving personnel and/ or material by clandestine means across a denied area or border. (JP 1-02)

reach back—The ability to exploit resources, capabilities, expertise, etc., not physically located in the theater or a joint operations area, when established. (MCRP 5-12C)

rear area—For any particular command, the area extending forward from its rear boundary to the rear of the area assigned to the next lower level of command. This area is provided primarily for the performance of support functions. (JP 1-02)

reconnaissance—A mission undertaken to obtain, by visual observation or other detection methods, information about the activities and resources of an enemy or potential enemy, or to secure data concerning the meteorological, hydrographic, or geographic characteristics of a particular area. (JP 1-02)

refugee—A civilian who, by reason of real or imagined danger, has left home to seek safety elsewhere. (JP 1-02)

repatriate—A person who returns to his or her country or citizenship, having left his or her native country, either against his or her will or as one of a group who left for reason of politics, religion, or other pertinent reasons. (JP 1-02)

Requirements Management System—A system for the management of theater and national imagery collection requirements. Provides automated tools for users in support of submission, review, and validation of imagery nominations as requirements to be tasked on national or DOD imagery collection, production, and exploitation resources. Also called RMS. (JP 1-02)

restricted area—**1.** An area (land, sea, or air) in which there are special restrictive measures employed to prevent or minimize interference between friendly forces. **2.** An area under military jurisdiction in which special security measures are employed to prevent unauthorized entry. (JP 1-02)

rules of engagement—Directives issued by competent military authority which delineate the circumstances and limitations under which US forces will initiate and/or continue combat engagement rules of engagement - with other forces encountered. Also called ROE. (JP 1-02)

sabotage—An act or acts with intent to injure, interfere with, or obstruct the national defense of a country by willfully injuring or destroying, or attempting to injure or destroy, any national defense or war material, premises or utilities, to include human and natural resources. (JP 1-02)

safe area—A designated area in hostile territory that offers the evader or escapee a reasonable chance of avoiding capture and of surviving until he can be evacuated. (JP 1-02)

safe haven—**1.** Designated area(s) to which noncombatants of the United States Government's responsibility, and commercial vehicles and material, may be evacuated during a domestic or other valid emergency. **2.** Temporary storage provided Department of Energy classified shipment transporters at Department of Defense facilities in order to assure safety and security of nuclear material and/or nonnuclear classified material. Also includes parking for commercial vehicles containing Class A or Class B explosives. (JP 1-02)

safe house—An innocent-appearing house or premises established by an organization for the purpose of conducting clandestine or covert activity in relative security. (JP 1-02)

sanitize—Revise a report or other document in such a fashion as to prevent identification of sources, or of the actual persons and places with which it is concerned, or of the means by which it was acquired. Usually involves deletion or substitution of names and other key details. (JP 1-02)

scientific and technical intelligence—The product resulting from the collection, evaluation, analysis, and interpretation of foreign scientific and technical information which covers: (a) foreign developments in basic and applied research and in applied engineering techniques; and (b) scientific and technical characteristics, capabilities, and limitations of all foreign military systems, weapons, weapon systems, and materiel, the research and development related thereto, and the production methods employed for their manufacture. (JP 1-02)

SECRET Internet Protocol Router Network—Worldwide SECRET level packet switch network that uses high-speed internet protocol routers and high-capacity Defense Information Systems Network circuitry. Also called SIPRNET. (JP 1-02)

security—**1.** Measures taken by a military unit, an activity or installation to protect itself against all acts designed to, or which may, impair its effectiveness. **2.** A condition that results from the establishment and maintenance of protective measures that ensure a state of inviolability from hostile acts or influences. **3.** With respect to classified matter, it is the condition that prevents unauthorized persons from having access to official information that is safeguarded in the interests of national security. (JP 1-02)

security classification—A category to which national security information and material is assigned to denote the degree of damage that unauthorized disclosure would cause to national defense or foreign relations of the United States and to denote the degree of protection required. There are three such categories:

a. Top secret—National security information or material which requires the highest degree of protection and the unauthorized disclosure of which could reasonably be expected to cause exceptionally grave damage to the national security. Examples of "exceptionally grave damage" include armed hostilities against the United States or its allies; disruption of foreign relations vitally affecting the national security; the compromise of vital national defense plans or complex cryptologic and communications intelligence systems; the revelation of sensitive intelligence operations; and the disclosure of scientific or technological developments vital to national security.

b. Secret—National security information or material which requires a substantial degree of protection and the unauthorized disclosure of which could reasonably be expected to cause serious damage to the national security. Examples of "serious damage" include disruption of foreign relations significantly affecting the national security; significant impairment of a program or policy directly related to the national security; revelation of significant military plans or intelligence operations; and compromise of significant scientific or technological developments relating to national security.

c. Confidential—National security information or material which requires protection and the unauthorized disclosure of which could reasonably be expected to cause damage to the national security. (JP 1-02)

security clearance—An administrative determination by competent authority that an individual is eligible, from a security standpoint, for access to classified information. (JP 1-02)

security countermeasures—Those protective activities required to prevent espionage, sabotage, theft, or unauthorized use of classified or controlled information, systems, or material of the Department of Defense. See also counterintelligence. (JP 2-01.2)

security intelligence—Intelligence on the identity, capabilities and intentions of hostile organizations or individuals who are or may be engaged in espionage, sabotage, subversion or terrorism. (JP 1-02)

sensitive—Requiring special protection from disclosure which could cause embarrassment, compromise, or threat to the security of the sponsoring power. May be applied to an agency, installation, person, position, document, material, or activity. (JP 1-02)

sensitive compartmented information—All information and materials bearing special community controls indicating restricted handling within present and future community intelligence collection programs and their end products for which community systems of compartmentation have been or will be formally established. (These controls are over and above the provisions of DOD 5200.1-R, *Information Security Program Regulation.*) Also called SCI. (JP 1-02)

sensitive compartmented information facility—An accredited area, room, group of rooms, or installation where sensitive compartmented information may be stored, used, discussed, and/or electronically processed. SCIF procedural and physical measures prevent the free access of persons unless they have been formally indoctrinated for the particular SCI authorized for use or storage within the SCIF. Also called SCIF. (JP 1-02)

sensor—An equipment which detects, and may indicate, and/or record objects and activities by means of energy or particles emitted, reflected, or modified by objects. (JP 1-02)

sensor data—Data derived from sensors whose primary mission is surveillance or target acquisition, such as air surveillance radars, counterbattery radars, and remote ground sensors. (MCRP 5-12C)

signal security—A generic term that includes both communications security and electronic security. Also called SIGSEC. (JP 1-02)

signals intelligence—**1.** A category of intelligence comprising either individually or in combination all communications intelligence, electronics intelligence, and foreign instrumentation signals intelligence, however transmitted. **2.** Intelligence derived from communications, electronics, and foreign instrumentation signals. Also called SIGINT. (JP 1-02)

situation assessment—Assessment produced by combining military geography, weather, and threat data to provide a comprehensive projection of the situation for the decisionmaker. (JP 1-02)

situational awareness—Knowledge and understanding of the current situation which promotes timely, relevant, and accurate assessment of friendly, enemy, and other operations within the battlespace in order to facilitate decisionmaking. An informational perspective and skill that foster an ability to determine quickly the context and relevance of events that are unfolding. Also called SA. (MCRP 5-12C)

source—**1.** A person, thing, or activity from which intelligence information is obtained. **2.** In clandestine activities, a person (agent), normally a foreign national, in the employ of an intelligence activity for intelligence purposes. **3.** In interrogation activities, any person who furnishes intelligence information, either with or without the knowledge that the information is being used for intelligence purposes. In this context, a controlled source is in the employment or under the control of the intelligence activity and knows that the information is to be used for intelligence purposes. An uncontrolled source is a voluntary contributor of information and may or may not know that the information is to be used for intelligence purposes. (JP 1-02)

special access program—A sensitive program, approved in writing by a head of agency with original top secret classification authority, which imposes need-to-know and access controls beyond those normally provided for access to confidential, secret, or top secret information. The level of controls is based on the criticality of the program and the assessed hostile intelligence threat. The program may be an acquisition program, an intelligence program, or an operations and support program. Also called SAP. (JP 1-02)

special activities—Activities conducted in support of national foreign policy objectives which are planned and executed so that the role of the US Government is not apparent or acknowledged publicly. They are also functions in support of such activities but are not intended to influence United States political processes, public opinion, policies, or media and do not include diplomatic activities or the collection and production of intelligence or related support functions. (JP 1-02)

special agent—A person, either United States military or civilian, who is a specialist in military security or the collection of intelligence or counterintelligence information. (JP 1-02)

special operations—Operations conducted by specially organized, trained, and equipped military and paramilitary forces to achieve military, political, economic, or informational objectives by unconventional military means in hostile, denied, or politically sensitive areas. These operations are conducted across the full range of military operations, independently or in coordination with operations of conventional, non-special operations forces. Political-military considerations frequently shape special operations, requiring clandestine, covert, or low visibility techniques and oversight at the national level. Special operations differ from conventional operations in degree of physical and political risk, operational techniques, mode of employment, independence from friendly support, and dependence on detailed operational intelligence and indigenous assets. Also called SO. (JP 1-02)

special purpose Marine air-ground task force—A Marine air-ground task force organized, trained and equipped with narrowly focused capabilities. It is designed to accomplish a specific mission, often of limited scope and duration. It may be any size, but normally it is a relatively small force—the size of a Marine expeditionary unit or smaller. It may contain other Service or foreign military forces assigned or attached to the Marine air-ground task force. Also called SPMAGTF. (Approved for inclusion in next version of MCRP 5-12C)

split base—Two or more portions of the same force conducting or supporting operations from separate physical locations. (MCRP 5-12C)

staff cognizance—The broad responsibility and authority over designated staff functions assigned to a general or executive staff officer (or their subordinate staff officers) in his area of primary interest. These responsibilities & authorities can range from coordination within the staff to the assignment or delegation to the staff officer by the commander to exercise his authority for a specified warfighting function or sub-function. Staff cognizance includes the responsibility for effective use of available resources and may include the authority for planning the employment of, organizing, assigning tasks, coordinating, and controlling forces for the accomplishment of assigned missions. Marine Corps orders and doctrine provide the notional staff cognizance for general or executive staff officers, which may be modified by the commander to meet his requirements. (Draft MCWP 6-2)

stay behind—Agent or agent organization established in a given country to be activated in the event of hostile overrun or other circumstances under which normal access would be denied. (JP 1-02)

strategic intelligence—Intelligence that is required for the formulation of military strategy, policy, and military plans and operations at national and theater levels. (JP 1-02) Strategic intelligence and tactical intelligence differ primarily in level of application but may also vary in terms of scope and detail.

strategic warning—A warning prior to the initiation of a threatening act. (JP 1-02)

subversion—Action designed to undermine the military, economic, psychological, or political strength or morale of a regime. (JP 1-02)

subversive activity—Anyone lending aid, comfort, and moral support to individuals, groups or organizations that advocate the overthrow of incumbent governments by force and violence is subversive and is engaged in subversive activity. All willful acts that are intended to be detrimental to the best interests of the government and that do not fall into the categories of treason, sedition, sabotage, or espionage will be placed in the category of subversive activity. (JP 1-02)

subversive political action—A planned series of activities designed to accomplish political objectives by influencing, dominating, or displacing individuals or groups who are so placed as to affect the decisions and actions of another government. (JP 1-02)

surveillance—The systematic observation of aerospace, surface or subsurface areas, places, persons, or things, by visual, aural, electronic, photographic, or other means. (JP 1-02)

surveillance and reconnaissance cell—Primary element responsible for the supervision of MAGTF intelligence collection operations. Directs, coordinates, and monitors intelligence collection operations conducted by organic, attached, and direct support collection assets. Also called SARC. (Change approved for inclusion in next version of MCRP 5-12C)

sustained operations ashore—The employment of Marine Corps forces on land for an extended duration. It can occur with or without sustainment from the sea. Also called SOA. (MCRP 5-12C)

tactical intelligence—Intelligence that is required for planning and conducting tactical operations. (JP 1-02) In Marine Corps usage, tactical intelligence is concerned primarily with the location, capabilities, and possible intentions of enemy units on the battlefield and with the tactical aspects of terrain and weather within the battlespace. (MCRP 5-12C)

tactical intelligence and related activities—Those activities outside the National Foreign Intelligence Program that: a. respond to operational commanders' tasking for time-sensitive information on foreign entities; b. respond to national intelligence community tasking of systems whose primary mission is support to operating forces; c. train personnel for intelligence duties; d. provide an intelligence reserve; or e. are devoted to research and development of intelligence or related capabilities. Specifically excluded are programs which are so closely integrated with a weapon system that their primary function is to provide immediate use targeting data. Also called TIARA. (JP 1-02)

tactical warning—**1.** A warning after initiation of a threatening or hostile act based on an evaluation of information from all available sources. **2.** In satellite and missile surveillance, a notification to operational command centers that a specific threat event is occurring. The component elements that describe threat events are: Country of origin—country or countries initiating hostilities. Event type and size—identification of the type of event and determination of the size or number of weapons. Country under attack-determined by observing trajectory of an object and predicting its impact point. Event time-time the hostile event occurred. Also called integrated tactical warning. (JP 1-02)

target—**1.** A geographical area, complex, or installation planned for capture or destruction by military forces. **2.** In intelligence usage, a country, area, installation, agency, or person against which intelligence operations are directed. **3.** An area designated and numbered for future firing. **4.** In gunfire support usage, an impact burst which hits the target. (JP 1-02)

target intelligence—Intelligence which portrays and locates the components of a target or target complex and indicates its vulnerability and relative importance. (JP 1-02)

task force counterintelligence coordinating authority (TFCICA)—The counterintelligence officer, or civilian equivalent, assigned responsibility for coordinating all counterintelligence activities within a joint task force. Also called TFCICA. The TFCICA has the authority to require consultation between the agencies involved, but does not have the authority to compel agreement. In the event that essential agreement cannot be obtained, the matter shall be referred to the appointing authority. Coordinating authority is a consultation relationship, not an authority through which command may be exercised. Together, the TFCICA and the DHS's HUMINT Operations Cell (HOC) form the nucleus of the J-2X. (JP 2-01.2) technical control - The performance of specialized or professional service, or the exercise of professional guidance or direction through the establishment of policies and procedures. (Proposed USMC definition per MCWP 6-2 and the next revision of MCRP 5-12C.)

technical surveillance countermeasures—Includes techniques and measures to detect and neutralize a wide variety of hostile penetration technologies that are used to obtain unauthorized access to classified and sensitive information. Technical penetrations include the employment of optical, electro-optical, electromagnetic, fluidics, and acoustic means, as the sensor and transmission medium, or the use of various types of stimulation or modification to equipment or building components for the direct or indirect transmission of information meant to be protected. Also called TSCM. (JP 1-02) technical survey -A complete electronic and physical inspection to ascertain that offices, conference rooms, war rooms, and other similar locations where classified information is discussed are free of monitoring systems. (JP 1-02)

telecommunication—Any transmission, emission, or reception of signs, signals, writings, images, sounds, or information of any nature by wire, radio, visual, or other electromagnetic systems. (JP 1-02)

tempest—An unclassified term referring to technical investigations for compromising emanations from electrically operated information processing equipment; these investigations are conducted in support of emanations and emissions security. (JP 1-02)

terrain intelligence—Processed information on the military significance of natural and manmade characteristics of an area. (JP 1-02)

terrorism—The unlawful use or threatened use of force or violence against individuals or property to coerce or intimidate governments or societies, often to achieve political, religious, or ideological objectives. (JP 1-02)

treason—Violation of the allegiance owed to one's sovereign or state; betrayal of one's country. (JP 1-02)

unconventional warfare—A broad spectrum of military and paramilitary operations, normally of long duration, predominantly conducted by indigenous or surrogate forces who are organized, trained, equipped, supported, and directed in varying degrees by an external source. It includes guerrilla warfare and other direct offensive, low visibility, covert, or clandestine operations, as well as the indirect activities of subversion, sabotage, intelligence activities, and evasion and escape. Also called UW. (JP 1-02)

unconventional warfare forces—United States forces having an existing unconventional warfare capability consisting of Army Special Forces and such Navy, Air Force, and Marine units as are assigned for these operations. (JP 1-02)

validation—A process normally associated with the collection of intelligence that provides official status to an identified requirement and confirms that the requirement is appropriate for a given collector and has not been previously satisfied. (JP 1-02)

warfighting functions—The six mutually supporting military activities integrated in the conduct of all military operations are:

1. command and control—The means by which a commander recognizes what needs to be done and sees to it that appropriate actions are taken.

2. maneuver—The movement of forces for the purpose of gaining an advantage over the enemy.

3. fires—Those means used to delay, disrupt, degrade, or destroy enemy capabilities, forces, or facilities as well as affect the enemy's will to fight.

4. intelligence—Knowledge about the enemy or the surrounding environment needed to support decisionmaking.

5. logistics—All activities required to move and sustain military forces.

6. force protection—Actions or efforts used to safeguard own centers of gravity while protecting, concealing, reducing, or eliminating friendly critical vulnerabilities. Also called WF. (MCRP 5-12C).

Warning—A communications and acknowledgment of dangers implicit in a wide spectrum of activities by potential opponents ranging from routine defense measures to substantial increases in readiness and force preparedness and to acts of terrorism or political, economic, or military provocation. (JP 1-02).

Appendix H. References

EO 12333 United States Intelligence Activities

NSCID 5 U.S. Clandestine Foreign Intelligence and Counterintelligence Abroad

Department of Defense Directives (DOD Dir)

0-2000.12	Combating Terrorism Program
1325.6	Guidelines for Handling Dissident and Protest Activities Among Members of the Armed Forces
3025.1	Use of Military Resources During Peacetime Civil Emergencies Within the United States, Its Territories and Possessions
	Human Resources Intelligence (HUMINT) Activities
5105.32	Defense Attaché System
5200.27	Acquisition of Information Concerning Persons and Organizations not Affiliated with the Department of Defense
S-5205.1	Acquisition and Reporting of Information Relating to National Security
5205.2	DOD Operations Security Program
5210.48	DOD Polygraph Program
5210.50	Unauthorized Disclosure of Classified Information to the Public
C-5230.23	Intelligence Disclosure Policy
5240.1	DOD Intelligence Activities
5240.2	DOD Counterintelligence Activities
5240.6	Counterintelligence Awareness Briefing Program
5525.5	DOD Cooperation with Civilian Law Enforcement Officials

Department of Defense Instruction (DODINST)

5210.84	Security of DOD personnel at U.S. Missions Abroad
5240.4	Reporting of Counterintelligence and Criminal Violations
5240.5	DOD Technical Surveillance Countermeasures (TSCM) Survey Program
C-5240.8	Security Classification Guide for Information Concerning the DOD Counterintelligence Program
S-5240.9	Support to Department of Defense Offensive Counterintelligence Operations
5240.10	DOD Counterintelligence Support to Unified and Specified Commands
5505.3	Initiation of Investigations by Military Criminal Investigative Organizations
5505.6	Investigation of Allegations Against Senior Officials of the DOD

Director of Central Intelligence Directives (DCIDs)

1/7	Security Control on the Dissemination of Intelligence Information
5/1	Espionage and Counterintelligence Abroad (With supplemental MOAs)

Defense Intelligence Agency Manual (DIAMs)

57-1	General Intelligence Production
57-6	DOD Indications and Warning System
58-1	Defense Intelligence Collection
58-7	Time Sensitive Requirements Coordination and Management
58-11	Department of Defense HUMINT Policies and Procedures
58-12	Department of Defense HUMINT Management System

Defense Intelligence Agency Regulation (DIAR)

DIA 60-4	Procedures Governing DIA Intelligence Activities that Affect U.S. Persons

Joint Publications (JPs)

	Concept for Future Joint Operations—Expanding Joint Vision 2010
0-2	Unified Action Armed Forces
1-02	Department of Defense Dictionary of Military and Associated Terms
2-0	Joint Doctrine for Intelligence Support to Operations
2-01	Joint Intelligence Support to Military Operations
2-02	National Intelligence Support to Joint Operations
3-02	Joint Doctrine for Amphibious Operations
3-07	Joint Doctrine for Military Operations Other than War
3-07.2	Joint Doctrine for Antiterrorism
3-10	Joint Doctrine for Rear Area Operations
3-13	Information Operations (with classified supplement)
3-13.1	Joint Doctrine for Command and Control Warfare
3-50.2	Doctrine for Joint Combat Search and Rescue
3-50.3	Joint Doctrine for Evasion and Recovery
3-54	Joint Doctrine for Operations Security
3-57	Doctrine for Joint Civil Affairs
5-00.2	Joint Task Force Planning, Guidance and Procedures
5-03.1	Joint Operations Planning and Execution System, Volume I
6-0	Doctrine for Command, Control, Communications and Computers Systems Support to Joint Operations

Secretary of the Navy Instructions (SECNAVINSTs)

3300.2	Combating Terrorism
3800.8B	Intelligence Oversight Within the Department of the Navy
S3810.5A	Management of Foreign Intelligence, Counterintelligence and Investigative Activities within the Department of the Navy
3820.2D	Investigative and Counterintelligence Collection and Retention Guidelines Pertaining to the Department of the Navy
3820.3D	Oversight of Intelligence Activities Within the Department of the Navy
3850.2B	Department of the Navy Counterintelligence
S3850.3	Support to Department of Defense Offensive Counterintelligence Operations
3875.1	Counterintelligence and Awareness Briefing Program
5500.30E	Reporting of Counterintelligence and Criminal Violations to Office of the Secretary of Defense Officials
5500.31A	Technical Surveillance Countermeasures (TSCM) Program
5500.34	Security of DOD Personnel at U.S. Missions Abroad
5520.3B	Criminal and Security Investigations and Related Activities Within the Department of the Navy

Chief of Naval Operations Instruction (OPNAVINST)

1620.1A	Guidelines for Handling Dissident and Protest Activities Among Members of the Armed Forces
3300.53	Navy Combating Terrorism Program
S3850.5	Support to DOD Offensive Counterintelligence Operations
C5500.46	Technical Surveillance Countermeasures
5510.1	Department of the Navy Information and Personnel Security Program Regulation

Army Regulation (ARs)

381-10	US Army Intelligence Activities
381-20	The Army Counterintelligence Program
381-47	US Army Counterespionage Activities
381-172	Counterintelligence Force Protection Source Operations and Low Level Source Operations

Marine Corps Doctrinal Publication (MCDPs)

3	Expeditionary Operations
5	Planning
6	Command and Control

Marine Corps Warfighting Publications (MCWPs)

0-1	Marine Corps Operations(Draft)
0-1.1	Componency
1	Warfighting
2	Intelligence
2-1	Intelligence Operations
2-11	MAGTF Intelligence Collections(Draft)
2-12	MAGTF Intelligence Analysis and Production(Draft)
2-13	MAGTF Intelligence Dissemination(Draft)
2-15.5	Interrogator-Translator Operations(Draft)
3-1	Ground Combat Operations(Draft)
3-2	Aviation Operations
4-1	Logistics Operations(Draft)
5-1	Marine Corps Planning Process
6-2	MAGTF Command and Control Operations(Draft)
6-22	Communications and Information Systems

Marine Corps Reference Publications (MCRPs)

4-27C	Enemy Prisoners of War and Civilian Internees
5-12C	Marine Corps Supplement to the DOD Dictionary of Military and Associated Terms

Marine Corps Order (MCOs)

3302.1	Antiterrorism Program
3850.1H	Policy and Guidance for Counterintelligence Activities
3820.1	Foreign Military Intelligence Collection Activities (FORMICA)
003850.2	Marine Corps Counterintelligence Force Protection Source Operations (CFSO)

Fleet Marine Force Publication (FMFRP)

3-23-2/FM 34-130	Intelligence Preparation of the Battlefield

Field Manuals (FMs)

34-5 (S)	Human Intelligence and Related Counterintelligence Activities (U)
34-60	Counterintelligence
34-60A	Counterintelligence Operations—Classified Supplement

Made in the USA
Monee, IL
04 November 2026

67387466R00171